The Origin of the Greek Alphabet:

A New Perspective

YAN Pui Chi   IP Lup-ng

The Commercial Press

The Origin of the Greek Alphabet: A New Perspective

Written by:	YAN Pui Chi (甄沛之) IP Lup-ng (葉立吾)
	Both authors contributed equally to this work.
Edited by:	Betty Wong
Cover designed by:	Cathy Chiu
Published by:	The Commercial Press (H.K.) Ltd.,
	8/F, Eastern Central Plaza, 3 Yiu Hing Road,
	Shau Kei Wan, Hong Kong
Distributed by:	The SUP Publishing Logistics (H.K.) Ltd.,
	16/F, Tsuen Wan Industrial Centre,
	220 –248 Texaco Road, Tsuen Wan,
	NT, Hong Kong
Printed by:	C & C Offset Printing Co. Ltd.,
	14/F, C & C Building, 36 Ting Lai Road,
	Tai Po, New Territories, Hong Kong

© 2022 The Commercial Press (H.K.) Ltd.

First Edition, First Printing, March 2022

ISBN 978 96207 0602 8

Printed in Hong Kong

Contents

In memory of my wife Louise Lam (藝瑛), who encouraged me to make a comparative study of the world's major writing systems. This work is a spin-off from the study.

YAN Pui Chi

Preface

The birth of the Greek alphabet in the first quarter of the first millennium BC marks a milestone in the development of the writing systems of mankind. A new type of writing system suited to the phonological structures of Greek and many other European languages had been invented by the Greeks at the latest by the eighth century BC. However, nobody has yet been able to give a satisfactory account of the full history of the birth of the Greek alphabet. The essence of the process by which this new writing system came into being is generally described as follows: The Greeks adopted the Phoenician alphabet to write their language. As the Phoenician writing system is consonantal, its signs denote only consonants. When some of the consonantal Phoenician signs changed into vowel letters, a new type of writing system was born. Some letters of the Greek alphabet denoted vowels, and the others consonants.

Back in 2008 we embarked on a comparative study of the world's major writing systems. When writing about the ancient scripts, we gradually came to realize that in the study of ancient scripts it is important to try to see them from the ancients' perspective in order to understand how the scripts actually work. With this in mind, we tried to study the Phoenician signs from the Phoenicians' perspective and came to realize that to regard the Phoenician writing system as consonantal has in effect prevented one from getting a true picture of the history of the birth of the Greek alphabet, because the Phoenician script does not operate on a consonantal basis. The Phoenicians' view of the nature of their script is actually quite different from the mainstream one held by most linguists today.

Linguists are generally agreed that a Phoenician sign stands for a consonant followed by any vowel or none. For example, the Phoenician sign ⌐ stands for the consonant /l/ followed by any vowel or none. As the Phoenician vowel system is generally assumed to comprise the following vowels: /a, i, u, aː, iː, uː, eː, oː/, the Phoenician sign ⌐ presumably stands for /la, li, lu, laː, liː, luː, leː, loː/ or simply /l/. This is the phonemic view of the nature of a Phoenician sign.

The mainstream view of the nature of a Phoenician sign even goes one step further than the phonemic one. It focuses on the unvarying initial consonant that a Phoenician sign represents while ignoring the variable vowel, if any, that follows the consonant. Linguists holding this view assert that Phoenician signs denote only consonants. For example, the Phoenician sign Ϲ denotes only the consonant /l/.

It is understandable that Western linguists describe the sound of a sign in a foreign script in terms that are most familiar to them—terms like consonants and vowels. However, it must be noted that these terms and the concepts they express are alien to ancient peoples, including Phoenicians. The Phoenicians most likely regarded the Phoenician signs in a more primitive way. To them, a Phoenician sign had several sounds, and it was the context of a piece of writing that would tell them which sound it was, just as the context of the following sentence would tell the English readers how the word *read* should be read: *The book I've just read is easy to read*. While words with the same spelling but different pronunciations are fairly rare in English writing, it is a very common phenomenon in the Phoenician script for a Phoenician sign to be read differently to convey different grammatical meanings, such as case and number. On seeing for example the Phoenician sign Ϲ, the Phoenicians would recognize this fairly pictorial sign as a sign for a shepherd's goad, whose Phoenician name is akin to Hebrew *lāmed* 'goad'. The initial part of the sign name *lā* would trigger off in their mind all the sounds that the sign Ϲ could represent, namely /la, li, lu, laː, liː, luː, leː, loː/ and /l/. To write any of these sounds, the Phoenicians would use the same sign Ϲ. They would regard each of these sounds as a single whole. Most likely, they were not even aware that each sound could be split into smaller parts.

The Phoenicians, however, might run into situations in which even the context of a piece of writing failed to determine the sound of a Phoenician sign, such as the correct reading of an unfamiliar foreign name. They would then be obliged to use a so-called *mater lectionis* 'mother of reading' after a Phoenician sign to help indicate which sound it was. How does a *mater* tell the readers how the preceding Phoenician sign should be read?

Strangely enough, the *modus operandi* of a *mater* in Semitic alphabetic writing does not seem to be very well understood today. It is often said that a *mater* is a vowel indicator. Sometimes a *mater* is even regarded as a vowel letter. However, how a *mater* actually functions remains unexplained.

From the study of ancient Egyptian writing, we came to know that the *matres lectionis* used in Semitic alphabetic writing most probably originated from this ancient script, and by working out how the *matres* were born in Egyptian writing, we came to know how they were used to help indicate the sound of a foreign name. We believe that if one knows how the *matres* came into being in Egyptian writing, one will know how they actually function as sound indicators, and that, with a good understanding of their function as sound indicators, one can give a satisfactory account of the birth of the Greek alphabet. To find out how the *matres* came into being and how they functioned in Egyptian writing, the readers can refer to §8 of the main text.

It is possible that the first Semitic alphabet was invented by the scribes of the Hyksos dynasty, which ruled the Nile Delta from the Egyptian city Avaris for more than 100 years from about 1650 BC. Avaris had a large population of Semitic immigrants, and the Hyksos rulers were probably Western Semites. As Western Semites had no writing of their own, the Hyksos scribes wrote in ancient Egyptian. Some scribes might have used their knowledge of Egyptian writing to invent the first Semitic alphabet, from which the Phoenician script is descended via proto-Canaanite writing. The knowledge of the first Semitic alphabet, including how *matres* are used to write foreign names, would then be passed on down the generations from users to learners among the Western Semites.

The Phoenicians are well-known seafaring traders. When doing business with other Mediterranean peoples, Greeks included, the Phoenicians probably needed to record some Greek names, such as the names of their trading partners and the ports of call, with the aid of *matres*. The Phoenicians might have attempted to record Greek names as early as the twelfth century BC. We believe that these attempts mark the beginning of Greek alphabetic writing.

Sooner or later it would come to the notice of the Greeks that not only could the Phoenicians write out their names with merely about twenty Phoenician signs but they could also read out the names easily, sometimes with remarkable accuracy. This might have motivated some Greeks to learn the Phoenician alphabet. There is evidence in the various regional scripts of ancient Greece to suggest that the Greeks learned the Phoenician alphabet by noting how the Phoenicians wrote a Greek name. We believe that the Phoenician way of writing Greek names would be determined by both the Phoenicians' perception of Greek sounds and their method of writing foreign names. The way the Phoenicians wrote out the sounds of Greek names would set the model from which the Greeks developed their own writing.

This book argues that if proto-Greek alphabetic writing developed from the above model, it would evolve quite naturally into the different regional scripts of ancient Greece in the second quarter of the first millennium BC. If it is true that proto-Greek alphabetic writing began when the Phoenicians tried to record Greek names, then the better one can understand how the Phoenicians perceived Greek sounds and how they wrote foreign names, the better one can reconstruct proto-Greek alphabetic writing. When reconstructing proto-Greek alphabetic writing, one should of course also take into account the various regional scripts of ancient Greece, because one needs to explain how the former could plausibly evolve into the latter.

How the Phoenicians would write the Greek sounds /ki/ and /ku/ in proto-Greek alphabetic writing can be used as an example to illustrate the above point. Judging from the regional scripts of ancient Greece, we believe that it is possible that under the influence of their mother tongue the Phoenicians would hear the Greek sounds /ki/ and /ku/ as the Phoenician sounds /ki/ and /qu/ respectively, because they would probably identify the Greek sounds /ki/ and /ku/ with the Phoenician sounds /ki/ and /qu/. To write the sounds /ki/ and /qu/ in a foreign name, the Phoenicians would use the forms ⌐⅄ and Ύφ, which should be read from right to left in accordance with the direction of writing in Phoenician. The second signs ⌐ and Ύ are *matres* used to help indicate the sounds of the first signs ⅄ and φ. When the Greeks adopted the Phoenician alphabet, they simply followed the Phoenicians' lead in writing the Greek sounds /ki/ and /ku/ as ⌐⅄ and Ύφ. This, we believe, could be the reason why the Greek sounds /ki/ and /ku/ were written as ⸝Ͱ and Ύφ in the regional scripts of ancient Greece. For the Phoenician way of writing the other Greek sounds and for a fuller history of the birth of the Greek alphabet, see the main text.

To better understand how the Phoenician writing actually works, we learned from an Egyptian the writing system of modern Arabic, which is descended from Phoenician through Aramaic. To find out how the Phoenicians might perceive some Greek sounds, we requested our Egyptian teacher to transcribe in Arabic letters some Cantonese syllables beginning with /s/, /h/, and /tʰ/. The findings are revealing, for they provide surprisingly simple answers to some much-debated questions about the origin of some written forms used in the regional scripts of ancient Greece.

Even though the Phoenician way of writing Greek names is not attested, this does not mean that there was no need for the Phoenicians to write Greek names. The proto-Canaanite and the Phoenician inscriptions belonging to the period from the twelfth to the eighth century BC are so few and they are used for such limited purposes that small wonder there is no inscriptional evidence for the Phoenician way of writing Greek names. This book argues that the Phoenicians needed to record Greek names with the aid of *matres* in their commercial contacts with the Greeks. We believe that one can still figure out how they would record Greek names by studying how foreign names are written in ancient Egyptian and in Semitic alphabetic scripts such as modern Arabic or Hebrew. To this end, we studied how foreign names are written in ancient Egyptian and in modern Arabic.

This book offers a whole new perspective on the history of the birth of the Greek alphabet. In the book, the current views on this issue and ours are compared. As this book attempts to explain how the concepts of *vowels* and *consonants* arose, it should be of great interest to linguists and phoneticians, especially those taking an interest in the world's writing systems. General readers who are curious about the genesis of the Greek alphabet are also likely to find the subject of the book interesting.

Introduction

This book aims to give a clear account of how ancient Greek alphabetic writing could naturally evolve into the world's first segmental writing system, in which vowel and consonant letters are used to represent vowels and consonants respectively.

Current mainstream view has it that Greek alphabetic writing originated from Phoenician consonantal writing. After learning the Phoenician alphabet from the Phoenicians, the Greeks were able to use the Phoenician signs to denote the Greek consonants. All that was wanting were the vowel signs for denoting the Greek vowels. When some of the Phoenician signs turned into vowel letters, Greek segmental writing was born. Some letters in the Greek alphabet denoted vowels, and others consonants.

The above view is questionable because Phoenician writing is in effect not consonantal. To the Phoenicians, a Phoenician sign had several sounds, which today can technically be described as several CV syllables with a common onset. So strictly speaking, a Phoenician sign is a syllabic sign with multiple sound values, not a consonant letter. The concept of *consonant* would be alien to the Phoenicians, who would not use their signs to denote consonants. The Phoenicians would write syllable by syllable, not consonant by consonant.

As a Phoenician sign had several sounds, the Phoenicians would be obliged to use a *mater* after it to denote the intended sound when writing an unfamiliar foreign name, such as a Greek name. When the Greeks exploited the Phoenician method of writing Greek names with *matres* to write the Greek language, an embryonic form of Greek alphabetic writing would emerge. This book will explain in great detail in §11 and §12 how this embryonic form would ultimately evolve into a fully-fledged Greek segmental writing system.

The mainstream view on the genesis of the Greek alphabet simply takes for granted that consonant letters already existed in ancient Semitic alphabetic writings. This book, however, strives to give a comprehensive account of how consonant and vowel letters came into being in ancient Greek alphabetic writings without assuming the prior existence of consonant letters.

1 The role of Phoenician writing in the creation of the Greek alphabet

Nobody knows for certain why, how, when, and where exactly Greek alphabetic writing began. But it is quite certain that the Greek alphabet originated from the Phoenician script. There are evidences that this is the case. The early Greek letters are very much like the Phoenician signs. The order of the letters in the Greek alphabet is basically the same as that of the signs in the Phoenician alphabet. The Greeks called their letters Phoenician signs. They even called their letters by the Phoenician names, such as *alpha* and *bēta*, even though these names did not carry any meaning in Greek, apart from referring to the letters themselves. No wonder it is commonly accepted that the Greeks learnt the alphabet from the Phoenicians. Scholars holding this view are many, including Gelb (1952:176), Jeffery (1961:1), Chao (1968:109), Higounet (1969:63), Gaur (1984:118), Sampson (1985:99), Naveh (1987:175), DeFrancis (1989:175), Coulmas (1989:162), Healey (1990:35), Powell (2009:230), and Gnanadesikan (2009:208).

As can be shown by the table below on page 13, there is a striking resemblance between the Phoenician signs and the early Greek letters, and the names of many Greek letters are derived from Phoenician. The order of the signs in the two alphabets is essentially the same.

As regards the shapes of the Phoenician signs and the early Greek letters, it can be said that the signs in the table are quite representative of the Phoenician signs written in Phoenicia in the tenth and ninth centuries BC and the early Greek letters written in Greece in the eighth and seventh centuries BC. It should be noted that during the said periods the Phoenician signs are quite uniform in shape whereas the Greek letters have variant forms in different dialect areas.

The ancient Greeks called the early Greek letters *phoinikeia grammata* 'Phoenician signs'. The Greek word *phoinikeia* is derived from *phoinix*, one meaning of which is 'red'. The Greeks called Phoenicia *phoinike* 'country of purple cloth' and the people coming from that country *phoinikes* 'Phoenicians'. By *phoinikes* they probably referred to the people selling purple cloth from Phoenicia, a territory which is roughly equivalent to today's Lebanon. The Phoenicians, however, called themselves Canaanites and their land Canaan, an area which in those days probably covered modern-day Lebanon, Israel, the Palestinian territories, the western part of Jordan, and southwestern Syria.

Table 1 A table comparing the Phoenician and the early Greek alphabets

Phoenician alphabet		Greek alphabet	
Sign	Name	Sign	Name
𐤀	'āleph	A	alpha
𐤁	bēth	𐌁	bēta
𐤂	gīmel	𐌂	gamma
𐤃	dāleth	𐌃	delta
𐤄	hē	𐌄	epsilon
𐤅	wāw	Y, 𐌅	wau, digamma
𐤆	zayin	I	zēta
𐤇	ḥēth	H	ēta
𐤈	ṭēth	⊗	thēta
𐤉	yōd	𐌉	iōta
𐤊	kaph	𐌊	kappa
𐤋	lāmed	𐌋	lambda
𐤌	mēm	𐌌	mu
𐤍	nūn	𐌍	nu
𐤎	sāmekh	𐌎	xi
𐤏	'ayin	O	omicron
𐤐	pē	𐌐	pi
𐤑	ṣādē	M *	san
𐤒	qōph	𐌒	qoppa
𐤓	rēš	𐌓	rhō
𐤔	šin	𐌔 *	sigma
𐤕	tāw	T	tau

* The signs M and 𐌔 are probably derived from the sign W. See §11.1.1.2.

2 Controversy over the nature of Phoenician signs

Since Greek letters were doubtless derived from Phoenician signs, it is important to ascertain the true nature of Phoenician signs before one can understand how Greek alphabetic writing started. A Phoenician sign is generally introduced in books on writing systems as a sign that stands for a consonant. For example, the Phoenician written word 𐤉𐤋𐤌 for 'king' (to be read from right to left in accordance with the direction of Phoenician writing) is generally transliterated in roman letters as <mlk>, which obviously cannot be easily pronounced. Thus the statement that a Phoenician sign stands for a consonant cannot be taken literally. As all genuine scripts are glottographic, they can be read out in the language they represent, and in this respect the Phoenician script is no exception. A comparative study of the ancient and modern Semitic scripts and their representations in other languages will reveal that a Phoenician sign, in fact, stands for a consonant followed by any vowel or none. For example, the Phoenician sign 𐤌 stands for the consonant /m/ followed by any vowel in the Phoenician vocalic system, or no vowel.

However, scholars are divided in their opinions about the nature of a Phoenician sign. Many say that it is a sign denoting a consonant only without indicating a vowel, while some contend that it is a syllabic sign. Scholars regarding Phoenician signs as consonantal signs include Diringer (1968:159), Higounet (1969:42), Gaur (1984:3838), Sampson (1985:77), DeFrancis (1989:150), Healey (1990:9), Daniels (1996:4), and Swiggers (1996:261). Scholars regarding Phoenician signs as syllabic signs include Gelb (1952:147-153), Chao (1968:109), Robins (1971:116), Havelock (1976:27), and Powell (2009:163-174). The debate has persisted for over half of a century. We believe that one possible way to resolve this long-standing controversy over the nature of Phoenician signs is to try to look at the signs from the Phoenicians' perspective.

3 Phoenician signs seen from the Phoenicians' perspective

If asked about the nature of their Phoenician signs, a Phoenician would probably explain in non-technical terms that a Phoenician sign had several "sounds". To illustrate his point, he might read out all the possible "sounds" of a certain sign, say, ᛉ. These "sounds" would be transcribed today as /na/, /ni/, /nu/, /naː/, /niː/, /nuː/, /neː/, and /noː/. Of these eight "sounds", three are short and five are long.[1] To write any of these eight "sounds", he would use the same sign ᛉ. When reading the sign ᛉ used in actual writing, he would know which of the eight "sounds" it stood for.

The multiple sounds of a Phoenician sign would not pose much of a problem for the Phoenician in reading because he would not read sign by sign, but would take in a group of signs that formed a meaningful unit at one time. He knew which sound of a sign to read in a written word once he understood the meaning of the word from its context. A sign used in actual writing is in fact unlike a sign taken out of its context: the former had only one sound while the latter had multiple ones. Writing Phoenician with Phoenician signs was even easier for a Phoenician. In writing the spoken words in his language, he would break them down into the smallest sounds possible and use a sign for each sound. For any of the eight related "sounds", he would use the same sign. To write all the possible sounds in Phoenician, he needed only twenty-two signs.

It must be noted that what the Phoenician meant by a "sound" above is, in today's linguistic parlance, a syllable. To a Phoenician, a syllable was the smallest analysable unit of sound in Phoenician speech (Powell 2009:171). A Phoenician sign, therefore, stood for several related syllables. Today, it can be said that a Phoenician sign stood for a set of CV syllables with a common onset followed by different rhymes. However, this description applies only to a Phoenician sign taken out of its context. In actual writing, it stood for one and only one syllable. Take for example the Phoenician sign ᛉ. It stood for /na/, /ni/, /nu/, /naː/, /niː/, /nuː/, /neː/, *or* /noː/ in actual writing. Each of these syllables begins with a common /n/ and ends with a different rhyme. Thus it can rightly be said that ᛉ is a syllabic sign, even though it has multiple sound values.

---

1 The three short "sounds" /na/, /ni/, and /nu/ are basic. The five long "sounds" are derived from the three basic "sounds" mainly as follows: /na/+/ʔa/ >/naː/; /ni/+/ji/ >/niː/; /nu/+/wu/ >/nuː/; /na/+/ji/ >/nai/>/neː/; /na/+/wu/ >/nau/>/noː/.

Some linguists hold that in a purely syllabic system a single symbol represents one syllable (Simpson 1994:5055). It might be argued on this ground that strictly speaking a Phoenician sign is not a syllabic sign because it represents several syllables. However, it should be noted that in a segmental writing system like English, it is quite common for a letter to have multiple sound values. Take for example the letter <a>. In the following sentence, each <*a*> represents a different vowel sound in a British accent called Received Pronunciation: 'The vill*a*ge school m*a*ster w*a*shed his f*a*ce *a*nd h*a*nds.' That <a> represents multiple sounds does not prevent it from being called a vowel letter.[2] By the same token, that a Phoenician sign represents several syllables should not prevent it from being called a syllabic sign either. As can be seen from the English sentence above, the presence of a sign with multiple sound values does not affect the normal operation of a writing system. It is in fact quite common for a sign to have multiple sound values in many writing systems.

Many linguists think that a Phoenician sign represents the consonantal value of a syllable while ignoring the vocalic element (Olson 2003:1028). If a modern linguist were to explain this view to a Phoenician, the Phoenician would not understand what he or she said at first. It would take the linguist a long time to make the Phoenician understand that the sounds of a Phoenician sign, say, ヶ, could be written as NA, NĀ, NI, NĪ, NU, NŪ, NĒ, and NŌ, and that the letter N denotes an unvarying consonant whereas the other letters denote different vowels. The linguist might say that the Phoenician ヶ sign could be regarded as the letter N, which denotes a consonant only without indicating a vowel. The Phoenician would probably think this way of looking at the Phoenician signs unnecessarily complicated because their writing did not operate in this way. The multiple sounds of a Phoenician sign did not affect in any important way how they used the sign in writing and in reading.

---

2 A vowel or consonant sound will be referred to simply as a vowel or consonant in this book, while a vowel or consonant letter will always be referred to as a vowel or consonant letter, not simply as a vowel or consonant.

4 The nature of a Phoenician sign

Today one can study directly how the modern Arabic and Hebrew scripts relate to their spoken languages and how their letters are pronounced in a piece of writing. Since the above scripts are descended from Phoenician writing through Aramaic without structural modification, the inner structure of all these Semitic alphabetic scripts should be the same, just as the inner structure of the Greek, Etruscan, Latin, and English alphabetic scripts is the same. A Phoenician sign should thus be of the same nature as an Arabic or Hebrew letter: it is basically a syllabic sign with multiple sound values. The written form كتب of the Arabic word /kataba/ 'he wrote' can be used as an example to illustrate the nature of an Arabic sign. It comprises three signs: ب + ت + ك, written from right to left. ك stands for /ka/, ت for /ta/, and ب for /ba/. When كتب means 'he wrote', the signs ك, ت, and ب are clearly syllabic. The same written form كتب can also be read as /kutub/, meaning 'books'. It can be seen from this example that the signs ك, ت, and ب have other sound values apart from /ka/, /ta/, and /ba/.

One might argue that ب is a consonantal sign when كتب is read as /kutub/, because ب stands for /b/. However, we contend that ب can still be regarded as a syllabic sign representing a syllable whose last part has weakened and finally disappeared. The /b/ in /kutub/ can be regarded as originating from a weakened syllable: /ba/, /bi/ or /bu/. This syllable has become so weakened and short that it is now ordinarily spoken as [b]. If this way of analysing the /b/ of /kutub/ can be established, we can put it more generally by saying that the final consonant of a closed Arabic syllable can be regarded as originating from a short syllable whose last part has weakened. An Arabic letter can then be regarded as a sign representing several syllables which have a common basic structure that can be represented by $c\alpha$, where c stands for an unvarying consonant and α for a variable vowel. When α is short, it can become so weak and short that it can be regarded as non-existent. If all the Arabic letters are regarded as representing $c\alpha$, then the phonological structure of Arabic words can be analysed in terms of $c\alpha$ syllables only.

Our analysis of the Arabic syllable is based on the hypothesis that the Semitic CVC syllable can be analysed as deriving, historically or even prehistorically, from two syllables: CV + CV. The evolution from CVCV to CVC is a main form of development in the Semitic languages and is beneficial in many ways to the development of the Semitic languages as systems of communication. According to

Sáenz-Badillos, a professor of Hebrew, by 1365 BC the disappearance of case-ending vowels and the development of the verb pattern *qatala* occurred in some Canaanite dialects, and prior to the ninth century BC, the loss of word-final short vowels took place (1993:45). These are all cases of evolution from CVCV to CVC.[3] Take for example the evolution from *qatala* to *qatal*. The shift took place probably because *qatal* required less effort to speak than *qatala*. A language community is a kind of "phonetic laboratory". One factor affecting phonetic change is the constant tug-of-war between the speaker's tendency to make the least effort possible in communication and the listener's demand for intelligibility. The development of the verb pattern from *qatala* to *qatal* must have gone through a long period of experimentation to make sure that the effectiveness of communication was not adversely affected before the verb pattern settled on *qatal*. As a matter of fact, the shift from CVCV to CVC or vice versa is not uncommon in the Semitic languages. For example, the Arabic word meaning 'you are writing' can be spoken either as /taktubu/ or as /taktub/. The last CVCV /-tubu/ can become CVC /-tub/, or vice versa.

As Phoenician is a Canaanite dialect, its CVC structure can be said to have derived from CVCV^V (C^V stands for a weakened CV). A Phoenician sign stands generally for CV and occasionally for C^V. If a Phoenician sign can be seen as a sign representing several CV syllables, including those which have weakened to C^V, then the twenty-two Phoenician signs can easily write out all the "sounds" of the Phoenician language. To the Phoenicians, a Phoenician sign stood for several "sounds" which they felt were somehow related both semantically and phonologically. These "sounds", including those weakened syllables C^V, would be regarded as belonging to the same category. The weakened syllables C^V would not be regarded as belonging to a separate category, as all the CV syllables will have a chance to turn into C^V in their spoken language when V is a short vowel. That is to say, the short Phoenician syllables, say, /ba/, /bi/ and /bu/, might be read as [bɐ], [bɪ], and [bʊ] respectively, or generally as [bə] or [b] in everyday speech as the Phoenician language evolved. Table 2 below shows a phonetic analysis of the sound values of the Phoenician syllables /ba/, /bi/ and /bu/:

---

3 Examples of CVCV turning into CVC can also be found in English. The first two syllables of the words medicine and reference may be spoken as /med-/ and /ref-/ respectively, apart from /medɪ-/ and /refə-/.

**Table 2 A phonetic analysis of the sound values
of the Phoenician /ba/, /bi/ and /bu/**

Full sound		Weakened sound		Further weakened sound		Minimal sound
[ba]	•••	[bɐ]	•••	[bə]	•••	[b]
[bi]	•••	[bɪ]	•••	[bə]	•••	[b]
[bu]	•••	[bʊ]	•••	[bə]	•••	[b]

The Phoenician syllable /ba/ is here used as an example to illustrate how the Phoenicians would regard the sound values of their basic syllables. What is said about /ba/ applies to /bi/ and to /bu/. When the Phoenician syllable /ba/ is spoken, its sound can vary in length and in accentuation as the actual conversational situation demands. These sounds may range from a fully articulated [ba] through a series of weakened sounds to [bɐ] or [bə] and then to [b]. [bə] is [b] followed by a barely audible schwa [ə], and [b] is a minimal sound without which /ba/ will become mute.[4] The Phoenicians would probably regard this whole series of sounds as different realizations of the same "sound" /ba/, of which [ba] or [bɐ] was the norm.

While the Phoenicians probably regarded [ba], [bɐ], [bə], and [b] as one "sound", a conventional phonemic analysis may treat these sounds differently. [ba], [bɐ], and [bə] may be regarded as different realizations of the underlying syllable /ba/, whereas [b] may be seen as the phonetic realization of the phoneme or consonant /b/. It seems to make just as much sense for the Phoenicians to have regarded [bə] and [b] as one sound as it does for a linguist to regard [ba] and [bə] spoken in Phoenician as different realizations of /ba/. Indeed, to a Phoenician uninitiated in the segmental concept, the phonetic difference between [bə] and [b] might even be smaller than that between [ba] and [bə]. However, in a phonemic analysis, while [ba], [bɐ], and [bə] may be regarded as belonging to the same category, [bə] and [b] may be regarded as belonging to two different categories, as can be shown from Table 3 below.

---

4 A minimal sound such as [b] and [p] may not be very audible in isolation, but its presence will be made more salient by a preceding vowel. For example, in the Arabic word /kutub/ 'books', the [b] sound at the end of the word is brought out by the preceding [u].

**Table 3 How the sound values of the Phoenician /ba/,
/bi/ and /bu/ are regarded in a phonemic analysis**

How the full sound and the weakened sounds are regarded	How the minimal sound is regarded
[ba] ··· [bɐ] ··· [bə] regarded as a manifestation of /ba/	[b] regarded as a manifestation of /b/
[bi] ··· [bɪ] ··· [bə] regarded as a manifestation of /bi/	[b] regarded as a manifestation of /b/
[bu] ··· [bʊ] ··· [bə] regarded as a manifestation of /bu/	[b] regarded as a manifestation of /b/

While it is universally agreed that a Phoenician sign is a phonogram, views are divided as to whether it is a syllabic sign or a consonantal one. The Phoenicians would probably regard a Phoenician sign as representing several "sounds", which they felt were somehow semantically related. Take for example the sign ᚼ. To the Phoenicians, it represents /ba/, /bi/, /bu/, /ba:/, /bi:/, /bu:/, /be:/, *or* /bo:/. When it represents a syllable with a short vowel /ba/, /bi/, or /bu/, it might be read respectively as [bɐ], [bɪ], and [bʊ], or generally as a further weakened [bə] or [b]. There is no doubt that [bə] could still be regarded as a syllable. As for [b], it is a weakened form of [ba], [bi], or [bu], and so one could argue that the inner or underlying structure of [b] is the same as that of [ba], [bi], or [bu]. As the underlying structure of [ba], [bi], or [bu] is undoubtedly a syllable, one could argue that [b] is structurally or underlyingly a syllable, too. If this argument holds, the formula *ca* which we use for analysing the sounds of a Semitic alphabetic sign is structurally or underlyingly a set of CV syllables, some of which can become so weakened that their rhyme (or V) may be elided. Since a Phoenician sign stands for *ca*, we believe that it is in essence a syllabic sign.

Many scholars hold that a Phoenician sign stands for the consonantal element of a CV syllable. The vocalic element of the syllable can be ignored as the Phoenicians would know which vowel to use, if any. This view implies that the Phoenicians knew how to isolate the consonant from the syllable. We deem it very unlikely for the Phoenicians to have known how to do that as the concept of *consonant* did not arise until the Greek segmental writing system had come into being. The Phoenicians did not need to have the concepts of *consonant* and *vowel* before they could use their script to read and write. To the Phoenicians, a Phoenician sign had several sounds or readings. The proper reading of the sign in a written word would pose little difficulty for them once they recognised which word it was from the context. When they were unsure about which word was used, they could try out the likely sound combinations of the signs. A few attempts would probably enable them to know the lexical core meaning and then which word it was.

There are twenty-two signs in the Phoenician orthography, with which the Phoenicians wrote their language. For example, the Phoenician word for '*king*' is written as 𐤊𐤋𐤌, to be read from the right, as explained in §2. 𐤌 stands for /m_/, 𐤋 for /l_/, and 𐤊 for /k_/. (Here the phonetic symbol within the slashes represents the syllable onset, while the underscore represents the variable rhyme. In the basic syllable structure *ca* represented by a Phoenician sign, the onset *c* is an unvarying consonant while the rhyme *α* is a variable vowel that can be elided at times.) Whichever is the sound value of /m_/ in 𐤊𐤋𐤌, it is still written as 𐤌. The same goes for /l_/ and for /k_/. The word /m_ l_ k_/, depending on its number, gender, and case, can be read in a great number of ways. No matter how it is read, the word is still written as 𐤊𐤋𐤌. When used in isolation, 𐤊𐤋𐤌 can be read in different ways. Once a Phoenician understood its meaning from its context, he would naturally know how to read it, being a native speaker of Phoenician. The Phoenicians wrote their language essentially syllable by syllable, not consonant by consonant.

To sum up, there are basically three ways to perceive the nature of a Phoenician sign. The table below is a critical summary of these three approaches:

Table 4 Three different ways to perceive the nature of a Phoenician sign

Syllabic approach	Phonemic approach	Consonantal approach
A Phoenician sign basically stands for several syllables related by a common onset, including any syllable whose last part has weakened. The notion of a syllable seems to be universal to all peoples. This notion should be within the grasp of the Phoenicians, who would call a syllable simply a sound. To the Phoenicians, a Phoenician sign stood for several related sounds.	A Phoenician sign basically stands for a consonant followed by any vowel or none. This is a phonemic description of the nature of a Phoenician sign. However, this description involves the use of the concepts of *consonant* and *vowel*, which were alien to the Phoenicians. It should be noted that the Phoenician signs were created long before the concepts of *consonant* and *vowel* arose.	A Phoenician sign basically stands for a consonant. The vowel that follows the consonant can be ignored as the Phoenicians would automatically know which vowel to use in a context. This implies that the Phoenicians knew how to isolate the initial consonant from the sounds of a Phoenician sign. This approach is even more abstract than the phonemic one.

5 Pre-phonemic vs phonemic interpretations of a Phoenician syllable

A Phoenician sign represents in effect several CV syllables, including the weakened C^V. It should be noted here that a Phoenician CV syllable can be spoken with varying degrees of accentuation of V in different polysyllabic words. Phonetically V could take on any value from a fully expressed V to a completely suppressed V, with varying degrees of accentuation of V in between. However, a phonemic interpretation of the various sound values of V (including a very weak V) in a Phoenician CV syllable demands that one should decide whether the said syllable is a CV syllable or simply a C phoneme. A pre-phonemic interpretation, as distinct from a phonemic one, would probably regard a Phoenician CV syllable as a single integral sound unit, leaving the native speakers to take care of the various values of V in daily conversation.

To illustrate the above point, we now use the English word *medicine* as an example. This word can be pronounced as /medɪsɪn/ or as /medsɪn/. In the pronunciation of /medɪsɪn/, the second syllable can be realized as [dɪ] or [də]. When [ɪ] or [ə] gets shorter and lighter, it may become inaudible, and the pronunciation of the word will be transcribed as /medsɪn/. A pre-phonemic approach would treat [dɪ], [də] and [d] as different realizations of the same sound, whereas a phonemic approach will regard [dɪ] and [də] as different realizations of the syllable /dɪ/, but will regard [d] as the realization of the consonant /d/. Accordingly, the syllable /dɪ/ and the consonant /d/ will be regarded as two different sounds.

A phonemic writing system has to decide on a spelling that reflects either the pronunciation /medɪsɪn/ or /medsɪn/. Since the spelling of a word has to take into consideration its inner morphological structure as well as its surface phonological structure, the spelling *medicine* has prevailed as it takes care of both structures. *medi-* is the root for a number of related words such as *medic, medical, medicinal*, and *medication*. If the word *medicine* were spelled as **medcine* so as to reflect the pronunciation /medsɪn/, the root *medi-* would not be as clearly identifiable. In a segmental writing system, sometimes one has to decide whether the spelling of a word should give a bias to its underlying morphological structure or to its surface phonological one. However, in a syllabic writing system like Phoenician, one does not have to deal with this problem. The sign representing the syllable /di/ can be left intact even when the actual pronunciation has weakened to [dɪ], [də], or [d].

6 The reason why a Phoenician sign has multiple sound values

A Phoenician sign is a phonogram that has multiple sound values, which is quite different from what we commonly know of a phonogram. We usually think that a phonogram basically stands for one sound, such as an Akkadian syllabogram or a Japanese *kana*. To write such syllables as /na/, /ni/, and /nu/, a Phoenician would use one phonogram while a Japanese person would use three. A Phoenician phonogram has multiple sound values because it can ultimately be traced back to a logogram with multiple sound values that was used in ancient Egyptian writing. For example, the Phoenician phonogram *2* can be traced back to the logogram ⊂⊃ used in ancient Egyptian writing. One has to understand how a logogram operates in ancient Egyptian writing before one can understand why *2* has multiple sound values (this point will be dealt with at greater length in §7).

Ancient Egyptian writing is generally thought to have been influential in the creation of early West Semitic writings, which ultimately gave rise to Phoenician writing. It can be said that without ancient Egyptian writing, there would have been no Phoenician writing with twenty-two signs in the second millennium BC, and that without Phoenician writing, there would have been no Greek alphabetic writing in the first millennium BC. Hence it is important to understand the nature of ancient Egyptian writing before one can understand how the Greek alphabet came about.

7 The nature of ancient Egyptian writing

To understand the nature of Egyptian writing, one must first understand the nature of the Egyptian language because the two are closely related to each other. The ancient Egyptian language and the Semitic languages belong to the Afro-Asiatic language family. As explained in §4, the basic syllable structure of the Semitic languages is ca. The basic syllable structure of the ancient Egyptian language should also be the same. By adhering to this basic yet flexible syllable structure ca, native speakers of the Egyptian language could easily form derivative or inflectional words according to a simple principle. Our analysis of the syllable structure of ancient Egyptian in terms of ca is basically consistent with Loprieno's (1995:36). Generally speaking, the phonological structure of an ancient Egyptian word is either c_1a_1, or $c_1a_1c_2a_2$, or $c_1a_1c_2a_2c_3a_3$. As early as five thousand years ago, the ancient Egyptians were able to write their language word by word either in pictorial signs called hieroglyphs or in a cursive or linear form of hieroglyphic writing called hieratic.

7.1 The Egyptian logograms

Egyptian hieroglyphs are used in monumental inscriptions. They are classified by scholars broadly into three types: logograms, phonograms, and determinatives (Davies 1987:30; Coulmas 2003:173). A logogram generally represents a common Egyptian noun whose pronunciation may vary with its case (this point will be elaborated below). The noun that a logogram represents generally refers to a common object or a living thing that could often be seen in the Egyptians' daily lives. So an Egyptian logogram is a word sign, which is a conventionalized pictogram that represents a whole spoken word. It is a motivated sign because its shape is related to the sense of the spoken word that it stands for. For example, the logogram $\frown$, depicting a loaf, represents the spoken word for 'loaf' in Egyptian. As a spoken word has both sense and sound, a logogram created to stand for an Egyptian word perforce carries, besides its sense, its sound values. The logogram $\frown$ therefore carries not only the sense 'loaf' but also the sound of the Egyptian word it stands for, namely /t_/. It should be noted that this spoken word has multiple sound values even though it has the simple phonological structure of c_1a_1, and so when its logogram $\frown$ is used as a phonogram through rebus, it also carries the multiple sound values of the spoken word /t_/ 'loaf'. As a matter of fact, $\frown$ stands for several syllables related by an unvarying onset c_1 followed by variable rhymes α_1.

7.2 The Egyptian phonograms

Ancient Egyptian has a simple syllable structure ($c\alpha$) and only a few vowels. Linguists are generally agreed that there were six vowels in archaic or old Egyptian: /a/, /i/, /u/, /aː/, /iː/, /uː/ (Loprieno 1995:35). Another way to describe the Egyptian vocalic system is that it had three vowels: /a/, /i/, /u/, each of which could be either long or short (Davies 1987:37). As it is generally presumed that ancient Egyptian had 23 or 24 consonants, there were at most: 24 c × 6 α = 144 $c\alpha$ syllables in the language (excluding those weakened $c\alpha$ syllables in which α had almost disappeared). As was said earlier, the phonological structure of an ancient Egyptian word is generally $c_1\alpha_1$, $c_1\alpha_1c_2\alpha_2$, or $c_1\alpha_1c_2\alpha_2c_3\alpha_3$. Discounting those weakened $c\alpha$ syllables in which α had almost disappeared, the ancient Egyptians could, in theory, create at most: (a) 144 words with the phonological structure of $c_1\alpha_1$ (24 × 6 = 144); (b) 20,736 words with the phonological structure of $c_1\alpha_1c_2\alpha_2$ (24 × 6 × 24 × 6 = 20,736); (c) 2,985,984 words with the phonological structure of $c_1\alpha_1c_2\alpha_2c_3\alpha_3$ (24 × 6 × 24 × 6 × 24 × 6 = 2,985,984). Each of the words thus created would be different in pronunciation. But in practice the ancient Egyptians would not exhaust all the possible sound combinations in their language in creating words because there was no such need. In fact, among all the words created by the ancient Egyptians, some would turn out to be, quite naturally, homophonous, just as in any other language.

When the ancient Egyptians had difficulty in creating a sign to represent a word whose sense is too abstract to be easily represented by a pictogram, they would be obliged to resort to the use of the rebus principle by borrowing a logogram that was homophonous with its stem or root to represent it. In fact, as can be seen from how Egyptian words are written in the ancient Egyptian script, many logograms are often used as phonograms, including those logograms with the phonological structures of $c_1\alpha_1c_2\alpha_2$ and $c_1\alpha_1c_2\alpha_2c_3\alpha_3$. It goes without saying, however, that logograms with the phonological structure of $c_1\alpha_1$ are used the most frequently as phonograms because phonograms of this kind are the most convenient, flexible, and versatile.

Approximately 750 logograms were used in the second millennium BC for the classical language of ancient Egyptian (Loprieno 1995:12). About 170 of them could also be used through rebus as phonograms to represent the sounds of words with identical stems or roots. For example, the three logograms ⌒, 𓄿, and ♀, originally used to represent the words /t_/ 'loaf', /m_s_/ 'fox fur', and /ʕ_n_x_/ 'sandal strap' respectively, were also used as phonograms to represent words or morphemes with identical stems or roots such as /t_/ 'feminine gender', /m_s_/ 'to give birth to', and /ʕ_n_x_/ 'life'. Egyptian phonograms are generally classified into three types, according to the number of consonants their sounds have: monoconsonantal, biconsonantal, and triconsonantal (Davies 1987:32; Loprieno 1995:12). So when used as a phonogram, ⌒ /t_/ is monoconsonantal, 𓅓 /m_s_/ biconsonantal, and ☥ /ʕ_n_x_/ triconsonantal. Since a phonogram was borrowed from a logogram through rebus, the former would naturally have as many sound values as the latter. If an Egyptian logogram had multiple sound values, then the phonogram derived from it would have multiple sound values.

To call an Egyptian phonogram "consonantal" is to call attention to its consonantal sounds, which are the unvarying elements of a phonogram whichever way the phonogram is pronounced. By counting the number of these unvarying consonants, one can divide the Egyptian phonograms conveniently and neatly into three types. Today these three types of phonograms are conventionally transliterated as $<c_1>$, $<c_1c_2>$, and $<c_1c_2c_3>$, which obviously cannot be easily pronounced. For the phonograms to be pronounceable, the vowel after c, if any, has to be supplied by the reader according to the context in which the phonograms are found. A fuller form of transliteration would be $<c_1a_1>$, $<c_1a_1c_2a_2>$, and $<c_1a_1c_2a_2c_3a_3>$. A phonogram that stands for c_1a_1 here would be regarded by the ancient Egyptians as representing several "sounds", each of which was a single integral sound unit. However, this kind of phonogram is generally introduced today as a monoconsonantal sign, which in fact represents several CV syllables in which V could vary or even disappear at times.

Of the three types of Egyptian phonograms, the most noteworthy are the monoconsonantal signs. It is generally believed that the Egyptian monoconsonantal signs most probably inspired the creation of the first Semitic alphabet. Thus a good understanding of the Egyptian monoconsonantal signs is of paramount importance. The better we can understand the nature of this type of signs, the better we know how the first Semitic alphabet came into being.

7.2.1 Why has an Egyptian logogram multiple sound values?

As explained earlier, the logogram ⌒ for 'loaf' can be used as a phonogram through rebus. The sound values of the phonogram ⌒ are quite certain, as it most probably represents /t/ followed by any vowel in the Egyptian vocalic system (the notation we use in this book is /t_/). As ancient Egyptian has six vowels: /a/, /i/, /u/, /aː/, /iː/, /uː/, the phonogram ⌒ most probably represents /ta/, /ti/, /tu/, /taː/,

/tiː/, /tuː/.[5] What needs explaining is why the phonogram 𓏏 can assume these sound values.

A brief explanation is that the logogram 𓏏, from which the phonogram 𓏏 is borrowed through rebus, has multiple sound values. The logogram 𓏏 has multiple sound values because the noun it represents has multiple sound values. The Egyptian noun for 'loaf', we believe, is spoken as /taː/, /tiː/, or /tuː/. Each syllable is likely to be long rather than short, for it represents possibly a fusion of a lexical morpheme and a grammatical morpheme.[6] The lexical morpheme for 'loaf' is spoken possibly as /t_/, and the case morpheme for 'accusative', 'genitive', and 'nominative' is spoken possibly as /-a/, /-i/, or /-u/. When the lexical morpheme is fused with the succeeding case-ending vowel to form the Egyptian noun for 'loaf', the noun is spoken as /taː/, /tiː/, or /tuː/. Its pronunciation varies with its grammatical case in a sentence.[7] The logogram 𓏏 that represents the Egyptian noun for 'loaf' is thus read as /taː/, /tiː/, or /tuː/. When the sign 𓏏 is used through rebus as a phonogram, it can obviously represent /taː/, /tiː/, /tuː/. The method of rebus, however, allows the phonogram 𓏏 to represent a syllable in any spoken word that has the same onset and rhyme as the spoken word /t_/ 'loaf', irrespective of vowel length. That is to say, the phonogram 𓏏 can also represent /ta/, /ti/, /tu/, apart from /taː/, /tiː/, /tuː/, because /taː/, /tiː/, /tuː/ rhyme respectively with /ta/, /ti/, /tu/.

The Egyptian logogram 𓏏 for 'loaf' probably represents a masculine noun as it is never followed by the feminine gender marker. When it has no written suffix to signify its number, it should stand for a singular noun. If this singular Egyptian noun /t_/ 'loaf' had no grammatical case to signify, it would be spoken as a single syllable only, not three. The Egyptian logogram 𓏏 that stands for the noun /t_/ 'loaf' would then have a single reading, and when used as a phonogram through rebus, the phonogram 𓏏 would have a single reading too, not several. However, the phonogram 𓏏 does have several readings. This indirectly shows that the Egyptian logogram 𓏏, from which the phonogram 𓏏 is borrowed, has several readings. The logogram 𓏏 has several readings because the noun that 𓏏 represents has pronunciations that vary

---

5 It should be noted that the three basic Egyptian syllables /ta/, /ti/, and /tu/, when weakened, can be realized as [t]. See Table 2, which shows that the three basic Phoenician syllables /ba/, /bi/, and /bu/ can be realized as a weakened [b].

6 An Egyptian syllable with a long vowel could have evolved from the $c_1 a_1 c_2 a_2$ structure, in which c_2 is a so-called weak consonant and a_2 is a short vowel. When the weaker $c_2 a_2$ is assimilated to the stronger $c_1 a_1$, the two syllables coalesce into a syllable with a long vowel.

7 The Egyptian noun for 'loaf' is spoken as /taː/, /tiː/, and /tuː/, just as the Ugaritic and Akkadian noun for 'mouth' is spoken as /paː/ (accusative), /piː/ (genitive), and /puː/ (nominative). About the pronunciations of the Ugaritic and Akkadian noun for 'mouth', see Gordon (1965:57).

with its grammatical case. This is also the reason why we believe that an Egyptian noun is inflected for case, just like an Akkadian or Ugaritic noun. The final part of a noun and the case-ending vowel seem to have fused together in ancient Egyptian. A logogram that stands for a noun perforce carries the pronunciations that vary with the cases. The phonogram that is borrowed from a logogram therefore also has various pronunciations.

7.2.2 The creation of the Egyptian monoconsonantal signs

We know from the Egyptian script that the monoconsonantal signs 𓅆 and ◠ are generally used after a logogram to write respectively the plural number and the feminine gender of an Egyptian noun. When a logogram is not followed by these signs, it can be assumed that the logogram probably represents a singular and masculine noun. The way in which the logogram ◠ is written and pronounced indicates that it probably stands for a singular and masculine noun. We believe that the ancient Egyptians would search among the logograms for those which represent a singular and masculine noun that is pronounced as *ca*, because these logograms could then be used as phonograms through rebus in a straightforward manner. Judging from the list of Egyptian monoconsonantal signs in Figure 80 of G. R. Driver's *Semitic Writing* (1948:135), the ancient Egyptians seem to have found six such signs in this way: 𓃀 /ʔ_/, 𓏭 /j_/, 𓂝 /ʕ_/, 𓉔 /h_/, 𓋴 /z_/, ◠ /t_/. Each sign represents six syllables, as explained at the beginning of this section. It seems that there are not many monoconsonantal nouns that refer to common objects or living things in the Egyptian language.

How would the ancient Egyptians write the other *ca* syllables in their language, such as the syllable for the preposition meaning 'towards'? We know from the Egyptian script that this preposition begins with /r/ and ends with a vowel. As prepositions are most unlikely to be inflected in their pronunciations, we believe that the Egyptian preposition for 'towards' is read as /r/ followed by a definite vowel. The ancient Egyptians would look for a suitable logogram that could be used as a phonogram through rebus to represent this sound. If they had difficulty in finding a monoconsonantal logogram for this purpose, they would resort to a biconsonantal or even triconsonantal one. In theory, they could choose any logogram whose first syllable is the same as the pronunciation of the preposition 'towards' and use it as a phonogram to represent, apart from the pronunciation of the preposition 'towards', /r/ followed by any vowel, i.e. /r_/. But judging from the monoconsonantal phonograms that the Egyptians finally used, as listed in G. R. Driver's Figure 80 mentioned above, they seemed to prefer logograms that begin with a strong syllable and end with one or even two weak syllables. For example, they finally chose the logogram ⟨mouth⟩ for 'mouth' and used it as a phonogram for writing the preposition 'towards'. The logogram ⟨mouth⟩ is read as /r_ʔ_/ or as /r_j_/. The first syllable /r_/ is probably both prominent and fixed, and is perhaps the same as the sound of the preposition 'towards'. It is possible, however, that the Egyptian noun for 'mouth' can also be spoken informally simply as /r_/ when the weaker syllable /ʔ_/ or /j_/ is assimilated to the preceding stronger syllable /r_/ (see footnote 5). If that is so, the remaining or surviving /r_/ should be inflected for case. The logogram ⟨mouth⟩ that represents the surviving /r_/ can thus be read as /raː/, /riː/, /ruː/. The logogram ⟨mouth⟩ then has multiple sound values, hence the phonogram ⟨mouth⟩. The ancient Egyptians might have found, in accordance with this principle, the following five monoconsonantal signs: ⟨mouth⟩ /r_/ < /r_ʔ_/ or /r_j_/, ⟨□⟩ /p_/ < /p_j_/, ⟨◉⟩ /x_/ < /x_j_/, ⟨pool⟩ /ʃ_/ < /ʃ_ʔ_j_/, ⟨hill⟩ /q_/ < /q_ʔ_ʔ_/.

However, two monoconsonantal signs ⟨hand⟩ and ⟨cobra⟩ seem to have originated from logograms in different ways. The hand logogram ⟨hand⟩ is read as /j_d_/. It is possible that the first syllable /j_/ is weak or unstressed whereas the second syllable /d_/ is strong or stressed. If this is the case, as the final part of a noun and the case-ending vowel are probably fused together in Egyptian, the noun for 'hand' may be read as /j_daː/, /j_diː/, or /j_duː/. Since the first syllable /j_/ is weak, it is possible that the noun for 'hand' can be spoken informally as /daː/, /diː/, or /duː/ when the meaning of this sound is clear in the context. Thus the logogram ⟨hand⟩ for 'hand' may also be read simply as /daː/, /diː/, or /duː/, depending on its case. The logogram ⟨hand⟩ can then be used through rebus as a phonogram with multiple sound values. How the cobra logogram ⟨cobra⟩ turned into a phonogram is more complicated, as it seems to represent a feminine noun whose gender needs to be indicated in writing as well. Four monoconsonantal signs ⟨cobra⟩, ⟨◟⟩, ⟨water⟩, and ⟨◔⟩ seem to have originated from logograms that represent feminine nouns. We now try to explain how these logograms came to be used as phonograms.

We start with the logogram ⟨water⟩ for 'water'. The Egyptian noun for 'water' is spoken as /n_t_/. One can regard this noun as consisting of the base /n_/, which conveys the lexical meaning 'water', plus the ending /t_/, which signifies 'the feminine gender'. The noun /n_t_/ 'water' is written as ⟨water⟩ on top of ⟨◠⟩. ⟨water⟩ is a logogram depicting rippling water, while ⟨◠⟩ is a monoconsonantal sign that indicates 'the feminine gender' /t_/. As ⟨◠⟩ represents the /t_/ in /n_t_/, the water logogram ⟨water⟩ can then be regarded as representing /n_/. In theory, the Egyptian case-ending vowels can be suffixed either to the noun base /n_/ or to the feminine gender marker /t_/. In the former case, the case-ending vowels will be fused together with the base. The /n_/ of /n_t_/ will probably be spoken as /naː/, /niː/, or /nuː/, depending on the case of the noun in a sentence. When the logogram ⟨water⟩ that represents /naː/, /niː/, or

/nu:/ is used as a phonogram through rebus, the phonogram ⁓⁓⁓ will obviously have multiple sound values.

However, judging from how case endings are suffixed to the feminine gender marker /t/ in both Akkadian and Ugaritic, we think it much more likely that the case-ending vowels in ancient Egyptian are likewise suffixed to and fused with the /t_/ sound for 'the feminine gender'. If that is so, the /t_/ in /n_t_/ will likely be inflected for case, and the sound of the base /n_/ will probably be both prominent and fixed, like that of the base of a noun in Akkadian and Ugaritic. In other words, the base /n_/ will probably be spoken as /n/ plus a definite vowel. The logogram ⁓⁓⁓ will then represent /n/ plus a definite vowel, i.e. a single CV syllable, and the phonogram derived from it will likewise represent a single CV syllable. This kind of phonogram that represents a single CV syllable will be distinct from the other phonograms that have multiple sound values, such as ⌒ , which basically represents six CV syllables. Four phonograms representing a single CV syllable seem to have derived from logograms in this way: ⁓⁓⁓ /nV/ < /nVt_/, ⌒ /fV/ < /fVt_/, ⊶ /çV/ < /çVt_/, ⌐ /ʤV/ < /w_ʔ_ʤVt_/.

We believe that the above four phonograms eventually changed from syllabograms with a single sound value into monoconsonantal signs with multiple sound values, possibly after a period of trial and error during which these two kinds of phonograms were in use to write different Egyptian words. The monoconsonantal signs with multiple sound values eventually prevailed probably because they are more appropriate for representing the Egyptian language than the syllabograms with a single sound value. To understand this point, one has to look into the morphological structure of a noun or a verb used in ancient Egyptian and Semitic languages. As was said earlier, the phonological structure of such a noun or verb is $c_1 a_1$, $c_1 a_1 c_2 a_2$, or $c_1 a_1 c_2 a_2 c_3 a_3$. Take for example a noun or a verb with the most common structure of $c_1 a_1 c_2 a_2 c_3 a_3$. It can be analysed as consisting of a lexical root $c_1 \_ c_2 \_ c_3 \_$, which is a sequence of three consonants used for conveying the lexical core meaning, and a pattern $\_ a_1 \_ a_2 \_ a_3$, which is a sequence of variable vowels interdigitating with the sequence of unvarying consonants to convey the grammatical or derived meanings. If monoconsonantal signs with multiple sound values are used to write such a noun or verb, it will be represented by an unvarying sequence of three signs, from which one can easily identify the skeletal structure of a written noun or verb and then grasp its lexical core meaning. However, if syllabograms with a single sound value, like Akkadian syllabograms or Japanese *kana*, are used to write such a noun or verb, it will be represented by a sequence of three syllabograms which may vary with the change of grammatical or derived meanings. As a result, the same noun or verb may be written in completely different combinations of syllabograms and so one cannot easily identify the written noun or verb from a glance at the sequence of the three syllabograms.

In ancient Egyptian writing, if the phonogram 〰 represented a single CV syllable, it would not be as versatile as a phonogram that represented several CV syllables with the same onset. One might wonder why the Egyptian scribes did not find another logogram that could be read as /na(:)/, /ni(:)/, /nu(:)/ to replace the phonogram 〰 . The answer might have something to do with the fact that this logogram was not easy to find. If that is so, the Egyptian scribes would be obliged to see whether they could use the phonogram 〰 as if it was a phonogram with multiple sound values, just like ⌓ , ⌢ , ⌤ , etc. If 〰 could actually work as a phonogram with multiple sound values, as it most likely could, they could then convert it into a phonogram with multiple sound values. The ancient Egyptians did end up using 〰 as a phonogram with multiple sound values. We believe that they might have found the following monoconsonantal signs in a similar way: 〰 /n_/, ⬻ /f_/, ➣ /ç_/, ⬎ /ʤ_/. Each sign stands for six syllables. The Egyptian method of creating these phonograms in the above way might have inspired the Semites to create their own alphabet in the first half of the second millennium BC. This point will be elaborated on when the creation of the Semitic alphabet is discussed.

According to G. R. Driver (1948:135), there is no way of telling precisely how some phonograms are derived from logograms through rebus, as the way in which these logograms are read remains unknown. Hence, the origin of some phonograms remains obscure. These phonograms are: 𓅱 /w_/, 𓃀 /b_/, 𓅓 /m_/, 𓉐 /h_/, 𓊽 /s_/, ⌣ /k_/, ⬚ /g_/, ⬌ /tʃ_/.

The Egyptian monoconsonantal signs seem to have arisen naturally from the evolution of the Egyptian script. To represent a monosyllabic function word whose sense is too abstract for easy graphic representation, the ancient Egyptians would be obliged to turn a logogram into a phonogram through rebus. They would search among the logograms for those which represent a noun with the $c_1\alpha_1$ phonological structure. They managed to find only a few such logograms, as the Egyptian nouns that refer to physical objects are generally disyllabic or polysyllabic. Thus the ancient Egyptians had to resort to those logograms that are characterized by having a strong syllable and one or two weak syllables. These weak syllables begin with the so-called weak consonants and so can easily be elided or assimilated. The Egyptians ended up using about twenty-five logograms as monoconsonantal signs. The different ways in which these signs came into being can be summarized in the table below:

Table 5 Different ways of inventing the Egyptian monoconsonantal signs*

Way of inventing the Egyptian monoconsonantal signs	Monoconsonantal signs thus invented
1. Phonogram derived through rebus from a logogram representing a noun pronounced as several CV syllables with a common onset	λ /ʔ_/, λ /j_/, — /ʕ_/, 口 /h_/, — /z_/, ᴐ /t_/
2. Phonogram derived through rebus from a logogram representing a noun with a strong syllable followed by one or two weak syllables	ᴐ /r_/, □ /p_/, ● /x_/, — /ʃ_/, ◿ /q_/
3. Phonogram derived through rebus from a logogram representing a noun with a weak syllable followed by a strong syllable	ᴐ /d_/
4. Phonogram derived through rebus from a logogram representing the first syllable of a feminine noun and then converted into a phonogram with multiple sound values	— /n_/, ↝ /f_/, ⬤ /ç_/, ˥ /ʤ_/
5. Unknown	ξ /w_/, ⌡ /b_/, λ /m_/, ↑ /h_/, ⎮ /s_/, ᴐ /k_/, ▣ /g_/, ⇒ / ʧ_/

* See Appendix 1 for a list of Egyptian monoconsonantal signs.

An Egyptian monoconsonantal sign is generally introduced as a phonogram that represents a consonant. It is often said, for example, that the phonogram ᴐ represents the consonant /t/. We believe that the concept of *consonant* would be alien to the ancient Egyptians. To them, the phonogram ᴐ had several concrete and pronounceable sounds, which can be transcribed today as /ta/, /ti/, /tu/, /taː/, /tiː/, /tuː/. When weakened, the Egyptian syllables /ta/, /ti/, /tu/ might be realized respectively as [tɐ], [tɪ], [tʊ], or generally as a further weakened [tə] or [t], just as the Phoenician syllables /ba/, /bi/, /bu/ might be realized respectively as [bɐ], [bɪ], [bʊ], or generally as a further weakened [bə] or [b] (see Table 2). In whichever way the above Egyptian syllables are pronounced, they can be represented by the same phonogram ᴐ . Even when they are pronounced as the minimal sound [t], they can be similarly represented. The phonogram ᴐ , in isolation, represents six full syllables, but in a piece of writing it represents either a full syllable or a weakened one. Hence the Egyptian monoconsonantal signs are in essence syllabic, not consonantal.

To put it more simply, the phonogram ᴐ can be regarded as representing three basic sounds /ta/, /ti/, and /tu/. These sounds can be lengthened or shortened. When lengthened, they become /ta:/, /ti:/, and /tu:/. When shortened, they become /t/.

Although it seems justified to call the Egyptian phonograms mono-, bi- and tri-consonantal signs, it should be noted that the Egyptian scribes would not see their phonograms as consonantal signs simply because they had no concept of *consonant*. They would regard each phonogram simply as a sign that could be pronounced in a number of ways. Normally they would not take the trouble to analyse the pronunciation of a phonogram. If they had to do so, they would probably break it up into syllables. To say that an Egyptian phonogram records only the unvarying consonant(s) while ignoring the variable vowels is to 'falsify essentially the Egyptians' own experience of writing in order to understand it' (Powell 2009:164).

To the Egyptians, it might be quite natural for a phonogram to have multiple sound values. Such phonograms would not pose much of a problem for the Egyptians, as they would not read an Egyptian word phonogram by phonogram, just as we would not read an English word letter by letter. When the Egyptians read, they would take in the shape of the word as a whole. Once they recognized a written word in a context, they, guided by their mother tongue, would automatically know how to read it.

7.2.3 How phonograms function in Egyptian writing

How did the Egyptians use their phonograms? To write the sound of a $c_1\alpha_1$ word, an Egyptian could use a monoconsonantal sign, as in the case of using the phonogram 𓂋 /r_/ for the word /r_/ 'to'. To write the sound of a $c_1\alpha_1c_2\alpha_2$ word, an Egyptian could use either two monoconsonantal signs, as in the case of using the phonograms 𓂋 /r_/ and 𓈖 /n_/ for the word /r_n_/ 'name', or a single biconsonantal sign, as in the case of using 𓄟 /m_s_/ for the word /m_s_/ 'to give birth to'. A monoconsonantal sign 𓋴 /s_/ was added to 𓄟 as a phonetic complement, not only to give an extra hint to the last part of the pronunciation of the word, but also to give the word a more distinctive written form. It should be noted that a phonetic complement is a sound indicator only and is silent in the written word. To write a $c_1\alpha_1c_2\alpha_2c_3\alpha_3$ word, an Egyptian could use either three monoconsonantal signs, or a monoconsonantal sign plus a biconsonantal sign, or a single triconsonantal sign. For example, to write the word /n_x_t_/ 'strong', the Egyptians conventionally used the monoconsonantal sign 𓈖 /n_/ plus the biconsonantal sign 𓆱 /x_ t_/.

7.3 The Egyptian determinatives

To use phonograms alone to represent words with the same phonological structure c_1a_1, $c_1a_1c_2a_2$, or $c_1a_1c_2a_2c_3a_3$ may not be sufficient to disambiguate their meanings. Thus sometimes there is a need to add determinatives to phonograms to clarify their meanings. For example, the Egyptian words meaning 'remain' and 'weak' are both pronounced /m_n_/, written as ⬜, ⬜ being the biconsonantal phonogram for /m_n_/ and 〰 being a complementary phonogram for /n_/. The Egyptians added the determinatives 𝄀 'book-roll' and and 🐦 'small bird' to clarify the meanings of the two words with the same phonological structure $c_1a_1c_2a_2$: ⬜ 𝄀 'remain' and ⬜ 🐦 'weak'. The book-roll signifies an abstract notion, whereas the small bird signifies something weak. It is worth noting that determinatives are semantic indicators only and do not carry any sound values.

7.4 The signs used in ancient Egyptian writing: a summary

To sum up, ancient Egyptian writing has three kinds of signs: logograms, phonograms, and determinatives. Logograms can be regarded as the basic signs because all phonograms and determinatives are derived from logograms. To write an Egyptian word, an Egyptian scribe had a large repertoire of signs at his disposal. He could use a single sign, which is either a logogram or a phonogram. Or he could use a combination of signs, which can vary in both kind and number. This enabled the Egyptians to create a rich variety of word forms that enhanced legibility. The pronunciation of each written noun or verb in ancient Egyptian varied with its grammatical meanings, while its written form might remain unchanged. However, this did not bother the Egyptians, because once they recognized its written form and grasped the meaning from its context, they knew how to read it. The reading of foreign names, however, was quite another matter because the context was of little help.

8 Writing foreign names in ancient Egyptian

An Egyptian name usually comprises a number of ordinary words. Writing an Egyptian name is no different from writing ordinary words, and so it also involves the use of logograms, phonograms, and determinatives. However, writing a foreign name is a different matter. The Egyptians would come to realize that the most direct and the easiest way of writing a foreign name was to use monoconsonantal phonograms to write its sound syllable by syllable, without bothering much about its sense. However, the problem was that they might not know how to read such a name afterwards, because even a monoconsonantal phonogram had multiple sound values. They had to find a way to specify the sound value of a monoconsonantal phonogram when writing a foreign name. This is where the so-called *matres lectionis* 'mothers of reading' came into play.

8.1 The birth of *matres lectionis* in ancient Egyptian writing

As their writing evolved, the Egyptians eventually discovered an ingenious way of writing the sounds of foreign names. As early as around 2150 BC, they came to know the use of three monoconsonantal signs 𓇋, 𓅱, and 𓄿 to aid the reading of foreign names (Loprieno 1995:14). Linguists call these signs *matres lectionis*. The Egyptians discovered that the phonograms 𓇋, 𓅱, and 𓄿 would lose their sound most easily when the following two conditions were met: (1) they were read lightly as /ji/, /wu/, and /ʔa/ respectively, and (2) they were preceded by a phonogram with the same rhyme.

Why would 𓇋, 𓅱, and 𓄿 lose their sound under these circumstances? Why would 𓇋, for example, lose its sound easily when it was read lightly as /ji/ and when it was preceded by a phonogram read as [ci] (c stands for any initial consonant)? One can explain the sound loss of 𓇋 from an articulatory standpoint. When [ji] is pronounced light and fast after a syllable ending in [i], there is a tendency for the two syllables to coalesce into a single syllable ending in [iː]. The sound changes can be shown like this: [ci] + [ji] > [ciji] > [ciii] > [ciː]. Since the sequence of sounds [i] + [j] + [i] in [ciji] is made by one and the same articulatory posture, and since the duration of [j] is much shorter than that of [i], the onset [j] is readily assimilated to the preceding peak [i]. The second syllable [ji], therefore, seems to have disappeared in the resulting syllable [ci:]. Thus, ⟨, representing /ji/, would lose its sound ultimately when read lightly and when preceded by a rhyming phonogram. By the same token, ⟨ would lose its sound easily when read lightly as /wu/ and when preceded by a rhyming phonogram. Since the relationship between [w] and [u] is similar to that between [j] and [i], the following sound changes would take place: [cu] + [wu] > [cuwu] > [cuuu] > [cu:]. The case of ⟨ is only slightly different. ⟨, representing /ʔa/, would also lose its sound easily when preceded by a rhyming phonogram. The sound loss of ⟨ can be explained in this way: [ca] + [ʔa] > [caʔa] > [caa] > [ca:]. The glottal stop [ʔ] tends to disappear when located between two [a] sounds because [aa] requires less effort to say than [aʔa].

⟨, ⟨, and ⟨ ultimately lost their sound in certain words under the circumstances described above. These three phonograms, despite their loss of sound, seemed to assume a new function of indicating the proper reading of the preceding phonogram. Take for example the Egyptian syllable /bi:/ that was written as ⟨⟨. This syllable was originally read as /biji/ before ⟨ lost its sound. The presence of the silent ⟨ thus seemed to indicate that ⟨ was to be read as /bi/, but not /ba/ or /bu/. By the same token, a silent ⟨ in ⟨⟨ would indicate that ⟨ was to be read as /bu/, but not /ba/ or /bi/. A silent ⟨ in ⟨⟨ would indicate that ⟨ was to be read as /ba/, but not /bi/ or /bu/. Over time, the Egyptians came to realize that ⟨, ⟨, and ⟨ could be used as auxiliary signs to indicate the proper reading of the preceding phonogram.

8.2 The creation of *matres* was need-driven

In a sense, a *mater* is like a determinative in ancient Egyptian writing. The creation of both the *matres* and the determinatives was probably driven by need. Since a monoconsonantal phonogram has three basic sound values, there is a need at times to use a *mater* to determine which of the three sound values is the intended one, as in the case of the writing of an unfamiliar foreign name. Just as a determinative is employed to disambiguate a written word with multiple semantic values, so a *mater* is used to disambiguate a phonogram with multiple sound values. A *mater* can thus be regarded as a sort of "phonetic determinative". To sum up, a *mater* and a determinative can be regarded as a sort of ancillary pointer, attached to the main body of a written word, thus enabling the reader to grasp the sound or meaning of the word more quickly. Many see a *mater* as a vowel letter (Healey 1994:74). This is not correct because a *mater* always represents a CV syllable in Egyptian writing, not a single vowel sound. A *mater* is not even a vowel indicator, as suggested by some linguists (Gelb 1952:169; Naveh 1987:9). Strictly speaking, a *mater* is a syllable indicator used to indicate the preceding phonogram's right syllable that rhymes with the *mater*.

8.3 The rhyming principle of *matres*

In many pre-literate societies rhymes were used as a powerful mnemonic device to help pass myths and legends orally from generation to generation. In ancient Egypt, however, rhymes were utilized for a new purpose, the significance of which is much greater than is generally realized. There, *matres* were used as rhymes to determine the sound values of phonograms in writing. Since ancient Egyptian has only three basic vowels or rhymes, the rhyming principle is very easy to use in ancient Egyptian writing.

The Egyptians would have little difficulty in reading *matres* in foreign names as *matres* were fairly easy to identify in Egyptian writing. When a reader encounters a foreign name in a piece of Egyptian writing, the context will point to its nature as such. Besides, the signs with which to write a foreign name always appear in a particular pattern in which each phonogram is generally followed by a *mater*. Since there are only three *matres*, their recurrence in a written word also makes it easy for a reader to recognize it as a foreign name. According to Gelb, there were intensive contacts in the second millennium BC between the Egyptian kingdom and the neighbouring lands. As there was a regular need to write foreign names in Egyptian, the convention to write these names systematically with *matres* was eventually established in ancient Egypt. To illustrate how the ancient Egyptians wrote foreign names with *matres*, Gelb cited the following examples (1952:168-169):

Table 6 Writing foreign names with *matres* in Egyptian

Foreign name	Transliteration of name written in Egyptian hieroglyphics*	Remark
Putu-Ḫipa	<p_*wu* t_*wu* x_*ji* p_*ʔa*>	Putu-Ḫipa is the name of a Hittite queen living in the 13th century BC. The name is written in Egyptian with four *matres*, each of which denotes the correct reading of the preceding sign. The spelling of this name with so many *matres* possibly reflects that the ancient Egyptians might not have been familiar with this foreign name and had to rely on the use of *matres* to get a definite reading of the name.
Ṣapūna or Ṣapōn	< ð_*ʔa* p_*wu* n_*ʔa* >	Ṣapūna is the name of a hill in Canaan. The three *matres* in the name enable the reader to get a definite reading of the name.
Carchemish or Karkamiša(š) or Karkamiš	<q_*ʔa* r_*ji* q_*ʔa* m_*ji* ʃ_*ʔa*> or <q_*ʔa* r_ q_*ʔa* m_ ʃ_*ʔa*>	Carchemish is the name of a Syrian city. In Egyptian it can be written with either five or three *matres*. In the latter case, the reader has to determine the readings of <r_> and <m_> himself.
Tunip	<t_*wu* n_*ji* p_*ʔa* ʔa> or <t_*wu* n_ p_>	Tunip was a Syrian city in the 14th century BC. Its name is written in Egyptian in two ways. The shorter name has only one *mater*, possibly because the hint of a single *mater* is enough for the reader to know which city is being referred to. The longer name with at least three *matres* enables an uninformed reader to read out the city name easily. The function of the last sign is little understood.
Kiz(zu)wat(a)na	<q_*ʔa* ð_*ʔa* w_*ʔa* d_*ʔa* n_*ʔa*>, <q_*ʔa* ð_ w_*ʔa* d_*ʔa* n_*ʔa*> or <q_*ji* ð_*ʔa* w_*ʔa* d_*ʔa* n_*ʔa*>	Kizzuwatna was an Anatolian kingdom in the 2nd millennium BC. The name is written with either four or five *matres*, which enable the ancient Egyptians to read it out easily.
Naharīna or Nahrīna	<n_*ʔa* h_ r_*ji* n_*ʔa*>, <n_ h_*ʔa* r_*ji* n_*ʔa*>, <n_ h_ r_*ji* n_*ʔa*> or <n_ h_ r_*ji* n_>	Naharīna or Nahrīna originates from the Syriac name for the Mesopotamia region. It is a name with four signs plus one to three *matres*. The fact that the name can be written with only one *mater* possibly suggests that the region was quite well known to the ancient Egyptians.

*A specific example can be used to illustrate how a foreign name written in Egyptian hieroglyphics is transliterated in this book. To write the sound /tu(ː)/ in a foreign name, the Egyptian scribes would use the so-called monoconsonantal sign ⌒ plus the *mater* 𓏲. ⌒ represents /t_/; 𓏲, a silent *mater* read as /wu/, requires the preceding sign ⌒ to rhyme with it. Thus ⌒ is to be read as /tu(ː)/. ⌒𓏲 are here transliterated into IPA symbols within angle brackets as <t_*wu*>. In the transliteration here the *mater* is written in italics. If <t_> is not followed by a *mater*, its reading is to be determined by the reader.

When reading the signs ⌒𓏲 in a foreign name, the Egyptian scribes would read them automatically as /tu(ː)/ as they were familiar with the use of *matres*. However, the Egyptian scribes would have to decide whether ⌒𓏲 should be read as /tu/ or as /tuː/. It must be stressed here that they would by no means see ⌒ as representing a consonant sound and 𓏲 as representing a vowel sound. It should also be noted that when writing ⌒𓏲 on papyrus, the Egyptian scribes would most probably use a cursive form called hieratic.

9 The first Semitic alphabetic script and some of its descendants

It is said at the outset of this book that nobody knows for certain why, how, when, and where exactly Greek alphabetic writing began. The same can also be said about the origin of Semitic alphabetic writing. The Proto-Sinaitic and the early Proto-Canaanite inscriptions are the earliest extant Semitic alphabetic writings, which may be dated to about 1700–1500 BC. These inscriptions attest to the Western Semites' first attempts to use the alphabetic signs to write their language. How did such signs come into being?

From a papyrus now in the Brooklyn Museum in New York, one can see how Semitic names were written by the ancient Egyptians. The papyrus, whose place of origin is probably Thebes in Egypt, contains a record of claims of ownership in a civil lawsuit between a noblewoman and her father. The record was written probably during the second half of the eighteenth century BC. It contains a long list of servants' names, at least 45 of which are identified as those of Asiatic people. These people, we believe, are most likely Semites coming from lands adjacent to the eastern part of Egypt. The name list points to both the possible presence of a sizable population of Semitic people in Thebes and the Egyptians' need to record their names in Egyptian hieroglyphs for various purposes. Such a need most likely arose earlier than 1700 BC, and the simplest way for the Egyptians to write Semitic names was probably to write their sounds by means of the Egyptian monoconsonantal signs.

9.1 The possible origin of the first Semitic alphabet

In regard to the writing of Semitic names by means of Egyptian monoconsonantal signs, one should perhaps mention the city of Avaris, which was situated at the eastern fringe of the Nile Delta and was connected to Sinai and Canaan by overland routes. It was established as early as the twentieth century BC by King Amenemhat I, who ruled Egypt from about 1985 to 1956 BC. It was called *Hutwaret* in Egyptian, which, when rendered in Greek, became *Avaris*. Because of its location Avaris was a popular destination for Semitic immigrants from nearby Canaan and other parts of Western Asia. By 1700 BC there was most likely a large population of Semitic immigrants in Avaris. In about 1650 BC the Hyksos dynasty set up court there. *Hyksos* means 'rulers of foreign lands'. Judging from the names of the six kings of the Hyksos dynasty, these foreign rulers were most probably Western Semites. They ruled the Nile Delta from Avaris for more than 100 years.[8]

Before the creation of the Semitic alphabet, the Western Semites did not know how to write their native language. The century-long Hyksos dynasty possibly provided conditions favourable for the creation of a Semitic alphabet. The few inscriptions left behind by the Hyksos reveal that the scribes wrote in Egyptian. If these scribes were descendants of the Semitic immigrants, as some of them most probably were, they were likely to be bilingual in Egyptian and West Semitic. Such scribes were in a good position to make an attempt to write their native language in hieroglyphs. As it is possible that the royal court and a large portion of the population in Avaris spoke West Semitic, these scribes might even have been assigned the task of finding a good way to write West Semitic. They must have known how the Egyptians wrote the Semitic names, and this way of writing Semitic names might have inspired them to create their own script. When reading a Semitic name written in hieroglyphs that contained, say, the monoconsonantal sign ᴧᴧᴧ , they would know that it represented the Semitic syllable /na/, /ni/, /nu/, /naː/, /niː/, /nuː/, /neː/, or /noː/ (assuming that West Semitic had these eight vowels at that time). In theory, they could use ᴧᴧᴧ to stand for these sound values too. However, as the sign ᴧᴧᴧ was pictorial, its meaning 'water' did not relate well to the above Semitic syllables. As the sign ᴧᴧᴧ was probably called *mēm* or the like in West Semitic, they would probably prefer to use the water sign ᴧᴧᴧ to stand for /ma/, /mi/, /mu/, /maː/, /miː/, /muː/, /meː/, or /moː/. The first CV syllable of *mēm* was /meː/. It was one of the syllables that the sign ᴧᴧᴧ could stand for, and it was a sound that could easily be associated with the object *water* designated by the sign. The use of the sign ᴧᴧᴧ to represent /m_/ would be more intimate to the Western Semites as it represented the first sound of the word *water* in their native language. They were thus inclined to use the sign ᴧᴧᴧ to represent /m_/ rather than /n_/.

The CV syllable /meː/ can be regarded as the acrophone of *mēm*, which is the Semitic name of the sign ᴧᴧᴧ . It should be stressed here that the acrophone of *mēm* is a CV syllable, not a consonant as held by many linguists. The acrophone of a Semitic noun is relatively constant and so can be associated more easily with the meaning of the noun than the succeeding sound of the noun. Take the Ugaritic noun for 'dog' as an example. It can be pronounced as /kalbu/, /kalbi/, /kalba/, /kalbatu/, /kalbati/, /kalbata/, etc. Its pronunciation varies with its number, gender, and case. No matter how it is inflected, its acrophone remains unchanged. It is always

---

8   See Mieroop (2011:126-145).

/ka/. Just as the sound /ka/ is easily associated with the meaning 'dog' in Ugaritc, so the sound /meː/ is easily associated by the Western Semites with the meaning 'water' and thus with the water sign 〰. When the sign 〰 represents /meː/, it is to a certain extent a motivated sign to the Semites, and this facilitates the learning of the Semitic alphabet. Once the Hyksos Semitic-speaking scribes had grasped the above acrophonic principle, they would be in a good position to create the necessary signs for representing all the syllables of their native language.

When adopting such Egyptian hieroglyphs as ⬓ , ⬯ , ⬎ , ⬰ , and ⬱ for use as Semitic alphabetic signs, the Hyksos scribes could just call these hieroglyphs by their Semitic name and then convert the signs that represented the acrophones of the designated nouns into signs with multiple sound values. After finding all the necessary signs, which should number less than thirty, they would probably try to arrange them in a certain order to form perhaps a kind of alphabet rhyme so as to facilitate learning. The first Semitic alphabet could have been formed in this way, and the Hyksos scribes in the Nile Delta area must have followed the Egyptian tradition of writing on papyrus. Given the climatic conditions in the Delta area, it is not surprising that the perishable papyri on which they wrote their native language can hardly be found today.

9.2 The Proto-Sinaitic and Proto-Canaanite alphabets

The creation of the first Semitic alphabet may not be as easy as it appears, but once created, it is not very difficult to learn. Given the chance and the motivation, even ordinary Semites can learn it within a short period of time. After the creation of the first Semitic alphabet, some Semites might have found it convenient to have a rudimentary knowledge of writing to meet some elementary needs in daily life, such as writing their own name, their gods' names, or some short wishes.

The idea that it was possible to use less than thirty pictographic signs to write Semitic words must have spread during the second quarter of the second millennium BC from Egypt eastwards to Sinai and Canaan. There are, however, only about thirty Proto-Sinaitic inscriptions and about twenty-five Proto-Canaanite inscriptions that can attest to the existence of the first Semitic alphabet that spread to Sinai and Canaan (Sass 1988:157). Moreover, the dates of the Proto-Sinaitic and early Proto-Canaanite inscriptions are uncertain. If it is true that the first Semitic alphabet was created during the Hyksos dynasty, then this alphabet could have spread to Sinai and Canaan in the seventeenth or sixteenth century BC. The Semitic alphabet that reached Sinai is called the Proto-Sinaitic alphabet, and the Semitic alphabet that reached Canaan is called the Proto-Canaanite alphabet.

From an inscription of a few linear signs $\sqrt{\upsilon \, \theta \, \gamma \, +}$ scratched on a stone sphinx in a temple in the Sinai Peninsula, one can see how the Western Semites put their writing to use. The inscription, dated to the second quarter of the second millennium BC, comprises five Proto-Sinaitic signs $\sqrt{\upsilon \, \theta \, \gamma \, +}$, read by the English Egyptologist Alan H. Gardiner as *lb 'lt* 'to (the goddess) *Ba'alat'*. We believe that these signs stand respectively for /l_/, /b_/, /ʕ_/, /l_/, and /t_/. The linear signs are still fairly pictorial. The first sign ⟋ looks like an ox-goad. Regardless of what it is called in the Egyptian language, the Western Semites called it by its Semitic name, which is akin to Hebrew *lāmed*. The sign, however, does not represent *lāmed* 'ox-goad'. It is a syllabic sign with multiple sound values. Not only can it represent the acrophone /la:/ of the noun *lāmed*, but it can also represent all the possible sounds with the same onset, namely /l_/. Here it stands for the Semitic word for 'to', which is pronounced as /l/ plus a definite vowel. The other Sinaitic signs are constructed most probably in the same way as ⟋. For example, the second sign ⎫ was probably derived from the Egyptian sign ⊏⊐, which depicts the floor plan of a house. In Egyptian, the sign ⊏⊐ , meaning 'house', was read as /p_r_/. The Western Semites adopted this sign and called it *bēth* or the like, which means 'house' in Semitic. The sign ⎫ represents /b_/.

The Proto-Sinaitic inscriptions are quite homogeneous. The signs are fairly pictographic. The stance of the signs is not fixed, and neither is the direction of writing. The signs can run vertically or horizontally. They can also run from right to left or from left to right. The Proto-Canaanite inscriptions are less homogeneous. The early signs are quite pictographic, while the later ones are more linear. It should be noted that the dates of the Proto-Canaanite inscriptions found in Canaan and Phoenicia range from the seventeenth century to the early eleventh century BC.

9.3 The Ugaritic alphabet and its significance

The first Semitic alphabet probably spread from Egypt in the seventeenth or sixteenth century BC eastwards to Sinai and Canaan, and then northwards to the eastern coast of the Mediterranean Sea, including the region called Phoenicia by the ancient Greeks. By the thirteenth century BC the Proto-Canaanite alphabet had reached, probably via Phoenicia, Ugarit, an ancient seaport to the north of the Phoenician coastal cities Tyre, Sidon, and Byblos. The use of a Semitic alphabetic writing system at Ugarit in the thirteenth and the early twelfth centuries is well attested by nearly two thousand clay tablets that have been unearthed from there since 1929.

Ugarit in the Late Bronze Age was a cosmopolitan city, where cuneiform was used to write such languages as Akkadian, Hurrian, and Hittite. Akkadian cuneiform was the lingua franca of the time used for international diplomacy and commerce in such areas as Mesopotamia, Anatolia, Canaan, and Egypt. The scribe or scribes who created the Ugaritic alphabetic cuneiform script must have been familiar with both the Akkadian cuneiform writing and the linear Semitic alphabetic writing that had spread to Ugarit. The Ugaritic scribes must have perceived the big advantage of the Semitic alphabetic script: it used far fewer signs than the Akkadian script. We believe that the scribes, inspired by the Semitic alphabetic writing, started to create their own script for writing Ugaritic, a Semitic language akin to Phoenician and Hebrew, while maintaining the Mesopotamian scribal tradition of writing cuneiform from left to right on clay tablets.

A clay tablet (KTU 1.6) inscribed in alphabetic Ugaritic contains a statement to the effect that the scribe Ilimilku was working under the patronage of King Niqmaddu. According to Bordreuil and Pardee, two experts on Ugaritic writing, the said King Niqmaddu should be Niqmaddu III (who died during the last decade of the thirteenth century BC), rather than Niqmaddu II (who died about 1350 BC) as was commonly believed. Their reason is that no tablets have been found with the names of the two kings Arḫalbu and Niqmepa (who reigned after Niqmaddu II) written in alphabetic Ugaritic. If the said King Niqmaddu had been Niqmaddu II, then, in all likelihood, the names of Arḫalbu and Niqmepa should have appeared on tablets inscribed in alphabetic Ugaritic. However, their names are not attested in alphabetic Ugaritic. This is probably because the Ugaritic alphabet had not yet been invented during their reign. This indirectly supports the hypothesis that the said King Niqmaddu should be Niqmaddu III (Bordreuil & Pardee 2009:19-20).

The above-mentioned clay tablet (KTU 1.6) thus attests to the fact that the Ugaritic alphabetic script was in use in the second half of the thirteenth century BC. The Ugaritic alphabet must have been invented at an earlier date. According to Bordreuil and Pardee, this alphabet had already been invented during the reign of Ammistamru II (ca. 1260–1235 BC), since his name is attested in Ugaritic alphabetic cuneiform whereas the names of Arḫalbu and Niqmepa (who reigned before Ammistamru II) are not. Thus they believe that the Ugaritic alphabet was not invented until sometime in the first half of the thirteenth century BC (2009:20). If their hypothesis is correct, then the Proto-Canaanite alphabet might have reached Ugarit in the fourteenth century BC. The Ugaritic scribes would need some time to familiarize themselves with the linear Proto-Canaanite alphabet before they could invent their own alphabet in cuneiform.

The Ugaritic alphabet is basically of the same nature as the West Semitic alphabets, such as Phoenician, Hebrew, and Aramaic, in which one sign represents several CV syllables with a common onset. It is unthinkable that this kind of alphabet could have been invented independently at Ugarit without an existing Semitic alphabet to base on.[9] Thus the prior existence of a Semitic alphabet is a prerequisite for the invention of the Ugaritic alphabet, and this Semitic alphabet should have been the Proto-Canaanite alphabet that had probably reached Ugarit in the fourteenth century BC.

The Ugaritic alphabet is well attested by more than a dozen abecedaries, the first of which was found in 1939 (Bordreuil & Pardee 2009:7). Abecedaries are lists of signs arranged in a standard order. The order of the signs in the Ugaritic alphabet is basically the same as in Modern Hebrew. It can be inferred from this that the above two alphabets should have derived from a common source, which is most likely the Proto-Canaanite alphabet. The Ugaritic alphabet originally has 27 signs, and the modern Hebrew alphabet has 22 signs. Despite the difference in the number of signs, the order of the signs in these two alphabets is essentially the same. It can be inferred from the original 27 signs in the Ugaritic alphabet that the Proto-Canaanite alphabet should have at least 27 signs. The 27 Ugaritic signs should be arranged essentially in the same order as in the Proto-Canaanite alphabet. There should also be abecedaries with 27 signs or more for the learning of the Proto-Canaanite alphabet, but their existence is not attested probably because they were written on papyrus or some other perishable material.

The task of inventing the Ugaritic alphabet would be made much easier for the Ugaritic scribes if they had for reference an abecedary of the Proto-Canaanite alphabet, which we believe they most probably had. They might have thought of borrowing the appropriate Akkadian syllabograms directly and then transforming them into signs with multiple sound values, but might have been deterred by the number of strokes or wedges required to write such syllabograms, which were a far cry from the simple signs in the Proto-Canaanite alphabet. The Proto-Canaanite signs written in the first half of the thirteenth century BC are barely attested by the extant Proto-Canaanite inscriptions, and so one can only guess that they should look slightly more pictorial than the better-attested twelfth-century Proto-Canaanite signs. The Ugaritic scribes probably had great difficulty in turning the still pictographic Proto-Canaanite signs into cuneiform ones. As they were very much used to writing unmotivated cuneiform signs, it is probable that they decided to design their own signs from scratch after much consideration. The form of each sign in the new script that they produced was simple yet distinct, and had on average much fewer strokes or wedges than the

---

9   In this sentence, 'this kind of alphabet' refers to any alphabet in which one sign represents several CV syllables with the same onset. In theory, it can be used to write any languages, including even non-Semitic languages, such as Hurrian and Greek. Nevertheless, how appropriate or effective it is for the languages concerned is another matter.

Akkadian syllabograms. With the invention of the new signs, the scribes could now use a small number of signs to write Ugaritic (see Appendix 2).

The early Proto-Canaanite signs can be said to be motivated because the name of a sign generally refers to the object depicted by the sign. How did the Ugaritic scribes call the signs in their alphabet, which apparently are not pictorial? They could have called the new signs by the Proto-Canaanite names. However, as the Ugaritic signs are not pictorial from the start, the scribes might have found it rather unnatural to call them by their traditional names. They might have called the signs simply by the first CV syllables of these names. A broken Ugaritic abecedary tablet (KTU 5.14) shows a list of twenty signs, each of which is annotated by an Akkadian syllabogram. This syllabogram might indicate either the name or a sound value of the annotated sign. For example, the second sign 𝄫 in the Ugaritic alphabet might be called *bēth* or the like as part of the tradition. Or it might be called *bē*, as annotated by the Akkadian syllabogram with the sound value of /be/.

There are three signs appended to the original alphabet with 27 signs: 𝈓, 𝈨, and 𝈊, numbered as signs 28, 29, and 30 respectively. Sign 30 is used, some scholars believe, to represent a Hurrian sound which is generally transliterated as ś. Sign 28 stands for /ʔi/, /ʔiː/, and /ʔeː/, whereas sign 29 stands for /ʔu/, /ʔuː/, and /ʔoː/. Before the addition of these two signs, sign 1 ⤜ originally stands for /ʔ_/, i.e., it stands for /ʔ/ plus any vowel or none. After the addition of signs 28 and 29, sign 1 now only stands for /ʔa/ and /ʔaː/. However, a problem arose when signs 28 and 29 were added: How should the sound /ʔ_/ be represented in a word when the vowel after /ʔ/ has become /ə/ or is elided? If the scribes knew the sound of the original vowel before it weakened into /ə/ or was elided, then they would know which of the three signs (signs 1, 28 and 29) should be used. But if they did not know the sound of the original vowel, they would have to decide arbitrarily which of the three signs was to be used by consensus. If signs 28 and 29 had not been added, such a problem would not have arisen. The addition of signs 28 and 29 to the Ugaritic writing system has in effect both advantages and disadvantages.

The ways in which the Ugaritic alphabet was used at Ugarit during the thirteenth century BC provide an invaluable insight into the ways in which the Proto-Canaanite alphabet could have been used in nearby Phoenicia during the same period. From the clay tablets unearthed from Ugarit, one knows that alphabetic cuneiform was used for writing the native language at Ugarit for administrative, economic, epistolary, religious, literary, and scholastic purposes while Akkadian was used as a lingua franca for legal and diplomatic purposes. One can imagine that the Proto-Canaanite script and Akkadian were probably used in more or less the same ways in Phoenicia during the same period.

9.4 The Phoenician alphabet

It can be inferred from the invention of the Ugaritic alphabet in the first half of the thirteenth century BC that a Semitic alphabet consisting of at least 27 signs must have been in use at least for some time at Ugarit before the invention of the Ugaritic alphabet. The probable candidate for this Semitic alphabet was most likely the Proto-Canaanite alphabet that had spread to Ugarit via Phoenicia. Thus it is likely that this Proto-Canaanite alphabet was in use in the second half of the fourteenth century BC in Phoenicia. Apart from the circumstantial evidence of the Ugaritic alphabet, there is no extant data, however, to support this point.

Judging from the wide use of the Ugaritic script at Ugarit in the thirteenth century BC, the Proto-Canaanite alphabet should have also been widely used for writing the Canaanite dialects spoken in Phoenicia in the same period. However, the extant inscriptions barely attest to the use of the Proto-Canaanite alphabet in that area then. There are in fact very few Proto-Canaanite inscriptions that can be dated to the thirteenth century BC. The earliest extant Proto-Canaanite abecedary attested by the ostracon from ʿIzbet Ṣarṭah in Canaan is dated to the twelfth century BC. This abecedary seems to attest a Semitic alphabet with only 22 signs, five signs less than the Ugaritic abecedary. The Canaanite inscriptions found in Phoenicia, dated as belonging to the period between the twelfth and eleventh centuries BC, also attest to the use of an alphabet with 22 signs. It seems that some signs in the Proto-Canaanite alphabet were not required to write the Canaanite dialects in ʿIzbet Ṣarṭah and in Phoenicia in the twelfth century BC.

The Phoenician alphabet has at least 5 signs less than the Proto-Canaanite alphabet that gave rise to the invention of the Ugaritic alphabet, probably because these signs were no longer needed for representing Phoenician in the last two centuries of the second millennium BC. When the place of articulation of the following five dental or velar fricatives /x, θ, ð, ðˤ, ɣ/ in Proto-Canaanite had shifted backwards in the oral cavity to merge with /ħ, ʃ, z, sˤ, ʕ/ respectively in Phoenician, the signs that had originally stood for /x_, θ_, ð_, ðˤ_, ɣ_/ became redundant. The redundant signs, except the sign W that stood for /θ_/, became obsolete. As for the sign W, when /θ/ had merged with /ʃ/, theoretically W should have been discarded just as the other four redundant signs. However, for some unknown reason, the Phoenicians used the sign W to replace the original sign for /ʃ_/. See Appendix 2.

The Proto-Canaanite alphabet that gave rise to the invention of the Ugaritic alphabet should be a long alphabet with at least 27 signs, not the shorter alphabet with 22 signs as used in Phoenician writing. If the Ugaritic alphabet had originated from the shorter alphabet, the signs in the Ugaritic alphabet would have been arranged as follows: the first 22 signs would have been arranged in the same order as in the Phoenician alphabet, and the remaining five signs would have been appended to the end of the shorter alphabet. But this is not the ordering of the Ugaritic alphabet.

The Phoenician script, which stabilized around 1100-1050 BC, is a direct descendant of Proto-Canaanite. The early Phoenician signs are less pictographic than the early Proto-Canaanite signs, but resemble the late Proto-Canaanite signs. They may look different from the Proto-Sinaitic and early Proto-Canaanite signs, but as sound symbols they are all of the same nature. Like a Proto-Sinaitic or Proto-Canaanite sign, each Phoenician sign stands for ca. It is a syllabic sign with multiple sound values. If it is true that the Phoenician language has eight vowels like Ugaritic, then it can be said that a Phoenician sign stands for eight syllables plus a reduced syllable.[10] Since there are 22 signs in the Phoenician alphabet, the Phoenician signs stand for 176 ca syllables ($22c \times 8a = 176$ ca syllables) plus 22 reduced syllables, i.e., 198 syllables ((22×8) $\sigma + 22\sigma = 198\sigma$). With these 22 signs, the Phoenicians could write all the syllables in their language.

The earliest extant extended Phoenician text is the inscription on Aḥiram's sarcophagus dated to around 1000 BC. It is already a mature piece of alphabetic Semitic writing. As was said earlier, the Proto-Canaanite alphabet should have been widely used in Phoenicia in the thirteenth century BC. There should be extended Proto-Canaanite texts written in Phoenicia then, but these texts have not been found. If they were written on papyrus or some other perishable material, they may have been lost to us forever.

---

10 The "reduced syllable" here refers to a consonant followed by a barely audible schwa. It originates from a CV syllable in which V is short. However, this "reduced syllable" is generally regarded as a consonant in a phonemic analysis. See Tables 2 & 3.

Did the Phoenicians use *matres*?

According to Naveh (1987:62), there were no *matres* in early Phoenician inscriptions before the eighth century BC. This is not surprising as such inscriptions were few in Phoenicia. The Phoenician script should have been much more widely used from the eleventh century to the eighth century BC than was attested by the extant scanty inscriptions. We believe that among the Semites, the Phoenicians must have had the greatest need for using *matres* to write foreign names as they had to make extensive trading contacts with the other Mediterranean peoples. Records of these names were most likely made on papyrus, as it was a most convenient writing material for roving traders. However, it is a perishable material which can hardly survive the passage of time in most climactic conditions, hence the absence of evidence that *matres* were used to write foreign names in early Phoenician.

We contend that the Phoenicians should have known how to use *matres* to write foreign names for two reasons. First, the inventors of the first Semitic alphabet must have had a good understanding of the ancient Egyptian writing system, including the use of *matres* to write foreign names, and so when the need to write foreign names occasionally arose, they would simply do as the Egyptians did. Such knowledge would be passed on from user to learner of the Semitic alphabet and from generation to generation. By the eleventh century BC, the Semites could have been using the method of writing foreign names with *matres* for several hundred years in Canaan, and this method must have formed part of the Phoenician writing system. The Phoenicians must have found this method indispensable for the writing of foreign names. Second, the genesis of the Greek segmental writing system indirectly supports the hypothesis that the Phoenicians used *matres* in their writing. One would have great difficulty in explaining the genesis of the Greek alphabet satisfactorily if no Phoenician *matres* had been available for use in the initial phase of its development (this point will later be elaborated on).

Despite the absence of concrete evidence for the use of *matres* to write foreign names in early Phoenician, one can still figure out how the Phoenicians would set about writing a foreign name by studying how foreign names are written in ancient Egyptian and in modern Semitic scripts like Arabic and Hebrew. As far as the line of descent of the Phoenician script is concerned, Egyptian is its predecessor while

Arabic and Hebrew are its successors. By studying the way in which foreign names from both ends are written, one can probably form an idea of how the Phoenicians would write a foreign name. We have already examined by means of a few specific examples in §8.3 the Egyptian way of writing foreign names. We are going to study some more examples in modern Arabic, which, we believe, is a suitable choice among the Semitic alphabetic writings as its method of writing foreign names is based on the use of three *matres* like that of Egyptian and presumably that of Phoenician too. Table 7 below shows how foreign names are written in Arabic. Please note that the transliterations here should be read from right to left in accordance with the direction of Arabic writing.

Table 7 Writing foreign names in Arabic

Name in English	Name in Arabic*	Arabic transliterated into IPA symbols	Remark
Rome	روما	<aʔ_m uw_r>	This name has two signs and two *matres*. The *matres* و and ا in the name not only enable the name to be read out easily, but they also make its written form much more distinct. Without the two *matres*, the name would be much less recognizable.
Sony	سوني	<ij_n uw_s>	This name has also two signs and two *matres*. The *matres* in the name here are و and ي. What is said above also applies to the name here.
Coca Cola	كوكاكو لا	<aʔ_l uw_k aʔ_k uw_k>	This name has four signs and four *matres*. The *matres* و and ا in the name make its pronunciation clear and its written form distinct.
Alibaba	علي بابا	<aʔ_b aʔ_b ij_l aʕ>	The name here is that of an e-commerce company, which originates from a famous character in Arabic literature. It is written in two words in Arabic. Without *matres*, both words would have two signs only. Thus *matres* have to be used to give them a distinct form. Once the words are recognized, they can be read out easily.
Athens	أثينا	<aʔ_n ij_θ aʔ>	This name has three signs and two *matres*. The *matres* in the name make its pronunciation /ʔaθina/ clear and its written form distinct.
Paris	باريس	<_s ij_r aʔ_b>	This name has three signs and two *matres*. The *matres* in the name make the pronunciations of the first two signs clear. Once the name is recognized, the reader will know how to read the last sign <s_>.
Moscow	موسكو	<uw_k _s uw_m>	This name has also three signs and two *matres*. The *matres* in the name make the pronunciations of the first and the last signs clear. The pronunciation of the medial sign <s_> is to be determined by the reader.
Crete	كريت	<_t ij_r _k>	This name has three signs and only one *mater*, which, in such a combination as in the name, probably give a sufficiently distinct form to the name to make it easily recognizable. The name is probably so familiar to the Arabic reader that the hint of one *mater* is enough for the name to be read out easily.
Berlin	برلين	<_n ij_l _r _b>	This name has four signs and only one *mater*. What is said above about the Arabic name *Crete* basically applies to this name too.
Sardinia	سردينيا	<_ʔ ij_n ij_d _r _s>	This name has five signs and two *matres*, which altogether give the name a distinct form. The pronunciations of the signs <d_> and <n_> are made clear by the two *matres*, whereas those of the other three signs are to be determined by the reader.

Washington	واشنطن ‹_n_ˤt_n_ʃ aʔ_w›	This name has five signs and only one *mater*. What is said above about the Arabic name *Berlin* basically applies to this name.
Beijing	بكين ‹_n *ij*_k_b›	This name has three signs and only one *mater*. What is said about *Berlin* basically applies to this name too. The Arabic name is transliterated from an older spelling than the modern *pinyin* form *Beijing*. Thus the Arabic name does not reflect the spelling *Beijing*.
Cyprus	قبرص ‹_ˤs_r_b_q›	This name has four signs but no *mater*, which means: (1) the four signs already give a sufficiently distinct form to the name to make it recognizable; (2) it is such a familiar place name that the Arabic writing system can treat it as if it were a native word, which is usually written without *matres*.
London	لندن ‹_n_d_n_l›	This name also has four signs but no *mater*. What is said above about the Arabic name *Cyprus* also applies here.
Cleopatra	كليوباترا ‹aʔ_r_t aʔ_b_w *ij*_l_k›	This name originates from a Greek name. The name in Arabic has six signs and three *matres*, which altogether give the name a distinct form. There are two consonant clusters in the Greek name, namely /kl/ and /tr/, each of which is represented by two signs in Arabic. The pronunciation of the first sign is to be determined by the reader, whereas that of the second sign is determined by the succeeding *mater*.
		‹k_› and ‹t_› in this name will be read as a consonant followed by a barely audible schwa. ‹w_› will be read as /wu/. When the name is read fast, the first four signs كليو, aided by the *mater* ‹*ji*›, will tend to be read as /kliu/.
Aristoteles	أريستوتلس ‹_s_l_t *uw*_t_s *ij*_r aʔ›	This name also originates from a Greek name. The name in Arabic has seven signs and two *matres*, which altogether give the name a distinct written form. The pronunciations of the signs ‹r_› and ‹t_› are made clear by the two *matres*, whereas those of the other five signs are to be determined by the reader. Those signs that are not followed by a *mater* can generally be read as unstressed sounds that carry a clear or an indistinct schwa.

* The name in Arabic is to be read from right to left.

Today the Arabic script is generally regarded as consonantal writing. Each Arabic sign or letter is taken to be a consonant letter. For example, in the name روما 'Rome', which is generally transliterated as <ruma>, the signs ر and م are seen as consonant letters that represent /r/ and /m/, whereas the *matres* و and ١ are seen as vowel letters that represent /u(:)/ and /a(:)/. However, to regard the Arabic writing as consonantal is questionable for two reasons. First, the sound of a *mater* in Arabic is not a vowel; it is always a CV syllable. The silent *matres* و and ١ should be read as /wu/ and /ʔa/, not /u/ and /a/. Second, native Arabic words are usually written without *matres*. If the Arabic signs were consonant letters, these words would not be easily pronounced. But the fact is that they are perfectly pronounceable. This means that an Arabic sign is not of the same nature as a consonant letter in a segmental writing system like English. All Arabic words, written with or without *matres*, are pronounceable. Even a foreign name written without *matres*, such as لندن 'London', is pronounceable. The letters <L, n, d, n> in the English word <London> are bona fide consonant letters whereas the signs <ن ,د ,ن ,ل> in the Arabic word <لندن> are not. An English consonant letter represents a sound that usually has to be brought out by sounding together with that of a vowel letter, while an Arabic sign represents several CV syllables which can be read out easily.

Foreign names written in Egyptian or Arabic are typically transliterations of names written in a foreign script. However, when the Phoenicians tried to write a Greek name in their trading contacts with the Greeks, they would not have a written form of the name to rely on. To record the name, they could only resort to the sound of the name actually spoken by native speakers. Owing to the nature of the Phoenician script, the Phoenicians would come to realize that the Greek names had better be written with the aid of *matres*. For example, if they wrote down the Greek name /nika:/ simply as 𝌀 (to be read from right to left) /n_ k_/, then they might not be able to recall how it should be read at a later time since 𝌀 could be read in a great number of ways. To give a definite reading to the name, they could simply follow the Egyptians' method of writing foreign names with *matres*.

The Phoenician signs 𝌀, ᴧ, and Y, we believe, can be used as *matres* to disambiguate the reading of a foreign name, apart from functioning normally like the other Phoenician signs. The Phoenician *matres*, we also believe, are only three in number because of their Egyptian origin. When used as *matres*, the signs 𝌀, ᴧ, and Y would be read as /ʔa/, /ji/, and /wu/ respectively. Since an ordinary Phoenician sign has multiple sound values, a *mater* is placed after it so as to specify its sound value in the writing of a foreign name. The *mater* requires the preceding sign to rhyme with it. It should be noted, however, that a *mater* is a syllable indicator only. It is a silent letter which does not represent any part of the preceding sign's sound. To write the Greek name /nika:/, the Phoenicians would use the most suitable *matres* after 𐤍 and 𐤊 to specify their sound value. How the Phoenicians would write a Greek name depends not only on their writing system, but also on their perception of the actual Greek pronunciation of the name. It goes without saying that the Phoenicians' perception of the Greek sounds was affected to a large extent by their mother tongue. The Phoenicians would probably write the name as 𐤀𐤊𐤉𐤍. Since 𐤍 /n_/ should rhyme with 𐤉 /ji/, and 𐤊 /k_/ with 𐤀 /ʔa/, the written name 𐤀𐤊𐤉𐤍 would be read as /nika:/. The Phoenicians should be happy with writing the Greek name /nika:/ as 𐤀𐤊𐤉𐤍, because to them there seemed to be little difference between the Greeks' pronunciation of the name and their way of reading the written name.

11 The Phoenician way to write a Greek name

As mentioned earlier, it is commonly agreed that the Greeks learnt the alphabet from the Phoenicians. However, despite this consensus about the provenance of the Greek alphabet from the Phoenician signs, scholars differ greatly on when and how the Greeks used the Phoenician signs to write Greek.

We believe that proto-Greek alphabetic writing began when the Phoenicians started to write Greek names in their commercial contacts with the Greeks. The Phoenicians exploited the sub-system of writing foreign names in their orthography to denote Greek sounds. When proto-Greek alphabetic writing started, its nature should be no different from that of this sub-system of Phoenician writing. However, the earliest extant Greek alphabetic writing, as inscribed on the Dipylon vase dated to around 740 BC, was already a mature piece of segmental writing. Nobody knows the actual processes by which proto-Greek alphabetic writing evolved into such mature segmental writing as was found on the Dipylon vase, because Greek alphabetic writings that antedate the Dipylon inscription have not yet been found.

When did proto-Greek alphabetic writing begin? Scholars suggest widely different dates, which vary from the 15th to the 8th century BC (Swiggers 1996:267). We believe that it might have taken more than a century for the Greek alphabetic writing to evolve from a syllabic system into a segmental one in the first quarter of the first millennium BC. Some historians of the classical era claimed that the Phoenicians founded the cities of Cadiz in Spain and Utica in Tunisia at the end of the twelfth century BC, but modern historians doubt the truthfulness of their claims, because there is scant archaeological evidence today to support these claims (Culican 1986:952). Despite the lack of physical evidence, one cannot rule out the possibility that the historians of the classical era had based their claims on records now lost to us. According to the modern archaeologist Maria E. Aubet, Phoenician cities like Byblos and Sidon soon recovered after the crisis of 1200 BC caused by the incursions of the Sea Peoples and resumed their commercial activities. At the end of the second millennium contacts between the Greeks and the Phoenicians were most intense (2001:29, 9). The Phoenicians are known to have a long history of seafaring and trading. Their overseas commerce with other lands might have been active during the latter half of the twelfth century BC. If this was the case, then the writing of Greek names by the Phoenicians might have taken place as early as the twelfth century BC when they came into contact with the Greeks en route to the Spanish coasts. In any event, the writing of Greek names by the Phoenicians should have happened no later than the end of the second millennium BC.

In their business transactions with the Greeks, the Phoenicians would need to write down some Greek names, such as the names of the ports of call, their trading partners, and even some produce indigenous to Greece. The Phoenicians might ask the Greeks to say these names slowly and clearly so that they could write them down syllable by syllable by means of Phoenician signs. When writing a Greek name, the Phoenicians would focus on its sound without bothering much about its meaning. As a Phoenician sign is a syllabic sign with multiple sound values, the Phoenicians would be obliged to use a *mater* to specify its sound value. They would choose the most suitable sign and *mater* to match as closely as possible each Greek syllable they heard. As there were only twenty-two Phoenician signs and three *matres* to choose from, the transcriptions might not be able to accurately reflect the Greek pronunciations.

The predominant syllable structure of Phoenician is CV. The CVC structure is also quite common. The Greek syllable structure, however, may comprise CV, CVC, V, and VC. The Phoenicians would not have much difficulty in writing CV syllables in Greek names, but would encounter some difficulty in writing other types of syllables. The written form of a Phoenician sign plus a *mater* specifically caters to the CV syllable, and so cannot be used conveniently to represent other types of syllable structure without modifications. As the sounds of foreign names need not be very exactly transcribed, the Phoenicians would probably tend to turn the other types of syllable structure into CV or CVCV structures when writing a foreign name. In the transcription of a foreign name, this way of turning its syllable structure to suit one's own can easily be found in scripts of such languages as Chinese and Japanese today, as these languages have a simple syllable structure like Phoenician's.

11.1 The Phoenician way to write a Greek CV syllable

To write a Greek CV syllable, the Phoenicians would use a Phoenician sign plus a *mater*. We first deal with the Phoenician signs used for writing Greek CV syllables in §11.1.1 and then with the *matres* in §11.1.2.

11.1.1 The Phoenician signs used for writing Greek CV syllables

The earliest extant alphabetic Greek inscriptions, which are dated to the second half of the eighth century BC, can be counted with the fingers of one hand. But in later centuries the number of alphabetic Greek inscriptions increased. By studying the numerous Greek inscriptions of, say, the fifth century BC, one can reconstruct with some certainty the phonological systems of the various Greek dialects spoken in that period. Since this book focuses on the time when the Greeks adopted the Phoenician alphabet, which we believe occurred probably in the tenth century BC, we have to hypothesize about the vowel and the consonant systems of the Greek dialects spoken then. Despite the lack of inscriptional evidence for the reconstruction of the phonological systems of the Greek dialects spoken in the tenth century BC, one may assume that they should not be drastically different from those in the fifth century BC. Besides, from the Phoenician signs that were first adopted for writing the Greek language, one can still form an idea about the consonant system of the Greek language spoken in the tenth century BC.

In ancient Greek, most dialects had in common fifteen consonants: /p^h, p, b, t^h, t, d, k^h, k, g, m, n, s, dz, l, r/ (Chadwick 1994:1494; Lejeune & Ruijgh 2003: 615). Eleven of them /p, b, t, d, k, g, m, n, s, l, r/ had close counterparts in Phoenician, and so Greek CV syllables beginning with these consonants can readily be represented by the corresponding Phoenician signs. For example, the Phoenician sign ◁, originally representing the Phoenician /b_/ syllables, would be ready for use to represent all the /b_/ syllables in Greek. Besides the above fifteen consonants, many Greek dialects also had a glottal fricative and a labialized velar approximant: /h/ and /w/. Greek syllables beginning with these two consonants could also be readily represented by the corresponding Phoenician signs ⟃ and Υ, which are meant for writing the Phoenician /h_/ and /w_/ syllables. Table 8 below shows the thirteen Phoenician signs that an outsider would expect the Phoenicians to use at the outset for writing Greek syllables beginning with consonants that had close counterparts in Phoenician:

Table 8 The Phoenician signs that one would expect the Phoenicians to use for writing the corresponding Greek CV syllables

Greek syllable	/p_/	/b_/	/t_/	/d_/	/k_/	/g_/	/m_/	/n_/	/s_/	/l_/	/r_/	/h_/	/w_/
Phoenician sign	⌐	◁	×	◁	⅄	∧	⁓	⁊	⧣	⟃	⟃	⧣	Υ
Phoenician syllable	/p_/	/b_/	/t_/	/d_/	/k_/	/g_/	/m_/	/n_/	/s_/	/l_/	/r_/	/h_/	/w_/

As can be seen from the local scripts of archaic Greece, nine of the above thirteen Phoenician signs did naturally evolve into the corresponding Greek consonant letters: ⌐ > ⊓/ Γ; ◁ > 8/ Β; × > Χ/ Τ; ◁ > Δ/ Δ/ Ð; ∧ > ⟍/ ⋏/ Γ/ Ϲ; ⁓ > Μ/

M; ꓢ > ꓥ / ꓠ / Ɲ ; ᑕ > ʟ / Ꮁ / ꓥ; ◁ > ◁ / ꟼ / R.[11] However, the representation of the Greek /k_/, /s_/, /h_/, /w_/ syllables by these four signs 〉, ╪, ⇁, and Y (highlighted in Table 8) met with some complications that need to be elaborated on below.

The representation of the Greek /k_/ syllables

How the Phoenicians would write the Greek /k_/ syllables is noteworthy. As can be seen from the Greek inscriptions written in the period from the eighth to the sixth century BC, both the letters Κ (which evolved from the Phoenician sign 〉) and Ϙ (which evolved from the Phoenician sign φ) were used to write the Greek consonant /k/. While the Greeks used Ϙ to write the onset of the Greek syllables /ku(:)/ and /ko(:)/, they never used it to write the onset of the Greek syllables /ki(:)/, /ke(:)/, and /ka(:)/. To write the onset of the Greek syllables /ki(:)/, /ke(:)/, and /ka(:)/, the Greeks always used the letter Κ. The use of Κ and Ϙ to write the Greek consonant /k/ can be seen as circumstantial evidence for the lead that the Phoenicians took in writing archaic Greek. In recording Greek names, the Phoenicians would use the sign 〉 for the Greek /ki(:)/, /ke(:)/, and /ka(:)/ as they would hear these syllables as sounds close to the Phoenician /ki(:)/, /ke:/, and /ka(:)/; however, they would use the sign φ for the Greek /ku(:)/ and /ko(:)/, which probably sounded to them like their /qu(:)/ and /qo:/. [k] is a velar stop articulated with the back of the tongue touching the soft palate, whereas [q] is a uvular stop articulated with the root of the tongue touching the very back of the soft palate. It is possible that the place of articulation for the ancient Greek /k/ varied to some extent according to the following vowel: it was further forward before a front vowel and further backward before a back vowel. According to Dionysius of Halicarnassus (c. 60 BC – after 7 BC), a Greek rhetorician, the Greek /k/ is said "with the tongue rising to the palate, near the pharynx" (Petrounias 2007:545). The place of articulation of the ancient Greek /k/ seems to be quite back. /k/ and /q/ are two different phonemes in Phoenician, while a front [k] and a back [k] are allophones in Greek. It seems that the Greek /k/ is realized as a sound close to the Phoenician /k/ when followed by a front vowel but close to the Phoenician /q/ when followed by a back vowel.

If the Greeks had had to write the Greek /ki(:)/, /ke(:)/, /ka(:)/, /ku(:)/ and /ko(:)/ on their own at the outset, they would have used the Phoenician 〉 only to represent all the /k_/ syllables, because to them these syllables all began with the same consonant /k/. They would not have used φ to write /ku(:)/ and /ko(:)/. The

---

11 It should be noted that when the Greeks wrote in boustrophedon style, the signs would appear in reversed forms. For example, the signs ꓶ and ꓭ would become Ꮁ and B when written boustrophedon.

Phoenicians' use of ⅄ and φ to write the Greek /k_/ syllables did not bother the Greeks at the beginning because they knew when to use ⅄ and when to use φ. However, when the logic of the phonemic principle asserted itself in Greek writing in the sixth century BC, the letter φ gradually fell out of use and was replaced by the letter Κ. /ku(:)/ and /ko(:)/ were almost universally written as ΚV and ΚΟ in Greece in the fifth century BC.

The representation of the Greek /s_/ syllables

The representation of the Greek /s_/ syllables seems to be more complicated than the case above. One would expect that under normal circumstances the Phoenician sign ⧫ sāmekh would be used to write the Greek /s_/ syllables because ⧫ represents the Phoenician /s_/ syllables. But in the local scripts of archaic Greece, the letter that was used to write the Greek /s/ was Ϟ / Ϛ sigma or Μ / Μ san, not ⧫. The Greek name sigma, however, vaguely suggests a link with the Phoenician name sāmekh. Some scholars believe that the name sigma was derived from sāmekh. If this is the case, it might imply that the Phoenician sign ⧫ sāmekh had once been used to write the Greek /s_/ syllables before it was abandoned by the mid-eighth century BC. We therefore think it possible that the sign ⧫ (sāmekh, later called by the Greeks sigma) had been used at the outset to write the Greek /s_/ syllables before it was replaced by the sign Ϟ / Ϛ sigma.

As regards the letter Μ san in the local scripts of archaic Greece, it suggests a stronger link with the Phoenician sign Ϣ shin with respect to both name and shape. Since Greek had only one sibilant: /s/, the Phoenician name shin would naturally become sin in Greek, which later evolved into san. With respect to shape, Μ and Ϣ apparently can be related. The sign Ϣ, when turned upside down, became Μ. There is evidence to suggest that Μ probably originated from Ϣ. In two local abecedaries of Corinth and Kroton dated c. 600-550 BC and c. 475-450? BC respectively (Jeffery 1961: Plate 20:16; Plate 50:19), the letter Μ san appears between <R> and <T>, just as the Phoenician sign Ϣ shin is located in the alphabet between ◁ rho and × tau. If it is true that Μ san originated from Ϣ shin, that means that san would naturally retain shin's position in the abecedaries and that Ϣ was probably used to write the Greek /s_/ syllables at one time. Why did the Phoenicians use Ϣ, which should stand for /ʃ_/, to write the Greek /s_/ syllables? The reason may be that some of the Greek /s_/ syllables sounded like their /ʃ_/ syllables. It is possible that the Phoenicians used both ⧫ and Ϣ to write the Greek /s_/ syllables.

Judging from the way in which the Cantonese /s_/ syllables are written in modern Arabic, we believe that it was possible for the Phoenicians to use not only ⧧ and ᗐ but also ↾ to write the Greek /s_/ syllables. It should be noted here that Cantonese, like archaic Greek, has only one sibilant phoneme, namely /s/. When requested to write out in Arabic letters the sounds of some Cantonese words read out to him, our teacher of Arabic, under the influence of his mother tongue Egyptian Arabic and constrained by the nature of the Arabic script, wrote the Cantonese syllables beginning with /s/ in the following three words as follows: 思/si/ سي, 書/sy/ شيو, 山/san/ صان. The Arabic transcriptions, if rendered in IPA symbols, would become: سي /siː/, شيو /ʃiːw/, صان /sˤaːn/. The highlighted Arabic letters س, ش, and ص represent /s_/, /ʃ_/, and /sˤ_/ respectively. It can be seen from here that the Cantonese phoneme /s/ was perceived as three different sounds /s/, /ʃ/ and /sˤ/ by our Arabic teacher. His perception of the Cantonese /s/ as three different sounds might be due to the fact that the sound of /s/ is affected by the vowel that follows it. The vowel /i/ in the Cantonese syllable /si/ is close, front, and unrounded like the vowel /iː/ in the English word *see*; the vowel /y/ in the Cantonese syllable /sy/ is close, front, and rounded like the vowel /yː/ in the German word *für* and the French word *sûr*; the vowel /a/ in the Cantonese syllable /san/ is open and back like the vowel /ɑː/ in the English word *father*. Hence our Arabic teacher used three different Arabic letters س, ش, and ص to transcribe the Cantonese /s_/ syllables.

The Arabic letters س, ش, and ص can be regarded as the respective counterparts of the Phoenician signs ⧧, ᗐ, and ↾: while ⧧ and س denote /s_/, ᗐ and ش denote /ʃ_/, and ↾ and ص denote /sˤ_/ (see Appendix 2). Just as the Arabic letters س, ش, and ص can be used to write the Cantonese /s_/ syllables, so the Phoenician signs ⧧, ᗐ, and ↾ could be used to write the Greek /s_/ syllables. If it is true that the Phoenicians used three different signs ⧧, ᗐ, and ↾ to write the Greek /s_/ syllables at the outset, it was probably the Greeks who decided to make the representation of /s_/ simple by using only one sign later on, after grasping the Phoenician alphabet. Judging from the local scripts of archaic Greece, the Greeks seemed to have chosen the sign ᗐ to represent /s_/.

Two other probable modern parallels of the Phoenicians' use of ⧧ and ᗐ to write the Greek /s_/ syllables can be found in the English speakers' use of <s_> and <sh_> to transcribe the /s_/ syllables in Japanese and Cantonese. The Japanese syllables /sa/, /su/, /se/, and /so/ are written as <sa>, <su>, <se>, and <so> respectively while /si/, /sja/, /sju/, and /sjo/ are written as <shi>, <sha>, <shu>, and <sho> in the Hepburn system, named after the American missionary James Curtis Hepburn. To Hepburn, the Japanese /si/ and /sj_/ sounded like the English /ʃi/ and /ʃ_/. Japanese scholars, however, uniformly wrote their /s/ phoneme with <s>, as shown by the Nihon-shiki and kunrei-shiki systems. Under these systems, the Japanese /si/ and /sj_/ are written as <si> and <sy_>. As regards Cantonese, the /s_/ syllables with a rounded front vowel are often written by the British as <sh_>. For example, the syllables /syt/ and /sœn/ are often written as <shuet> and <shun>. Even the syllables /sa/ and /san/ are often written as <sha> and <shan>, for unknown reasons. It is not very surprising, therefore, that the Phoenicians took some of the Greek /s_/ syllables for their /ʃ_/ syllables and so used W to write them.

The letter ⟩/⟨ *sigma* in the local scripts of archaic Greece, we believe, is a variant form of W (*shin* > *sin* > *san*), not a descendant of ⧧ (*sāmekh* > *sigma*). W , when turned 90° anti-clockwise, became ⟩, and then ⟨when written *boustrophedon*. ⟩ or ⟨, accordingly, should have been called *san* too. Why was it called *sigma* instead? One possible reason is that the name *sigma*, being commonly accepted as the name of the sign ⧧ for writing the /s_/ syllables, had persisted from force of habit even after the original sign ⧧ had been replaced by ⟩ .

As was said earlier, it is possible that the Phoenicians used three different signs ⧧, W , and ⱱ to write the Greek /s_/ syllables and that the Greeks decided later on to make their writing easier by using only one sign for this purpose. It is not very clear why W was preferred. One possible reason is that W was the easiest to write. The shape of ⱱ is not symmetrical enough for a beginner to learn it as easily as the other two signs ⧧ and W , which are more regular and symmetrical in shape. W was preferred to ⧧, possibly because W can be written in a single stroke without lifting the pen while ⧧ has to be written in four strokes.

The Greek historian Herodotus reported in the late fifth century BC that the same letter was called *san* by the Dorians and *sigma* by the Ionians (*Histories* 1.139). It is possible that by 'the same letter' Herodotus meant the letter ⟩/⟨, which was variously called *san* and *sigma* in different Greek regions. Regions that called ⟩/⟨ *san* might have come under the influence of those regions that called ⟩/⟨ *sigma* and so would gradually come to call ⟩/⟨ *sigma* too. The name *sigma* finally came to replace the name *san* and was adopted as the proper name for the letter ⟩/⟨. It should be noted that to the Phoenicians, the shape of a Phoenician sign, however abstract, could still be considered motivated because the name of a Phoenician sign still reminded them of the object that the name stood for, but to the Greeks, the relation between the shape of a Phoenician sign and its name was basically arbitrary. Thus the Greeks could freely associate the name *sigma* with the letter ⟩/⟨. Though called *sigma*, the letter ⟩/⟨ still retained *shin*'s position in the alphabet.

The representation of the Greek /h_/ syllables

One would expect the Phoenicians to use the sign ∃ *he* to write the Greek /h_/ syllables because ∃ represents the Phoenician /h_/ syllables. But in the local scripts of archaic Greece, the letter that was used to write the Greek /h/ was 日 / 目 / H *ḥēth*, not ∃ *he*. In Phoenician, while the sign ∃ is meant for writing the /h_/ syllables, the sign 日 is meant for writing the /ḥ_/ syllables. Why was 日 used to write the Greek /h_/ syllables instead of ∃?

Judging from the way in which the Cantonese /h_/ syllables are written in Arabic letters, it is possible that the Phoenicians used both the signs ∃ and 日 to write the Greek /h_/ syllables. Our Arabic teacher wrote in Arabic letters the following four Cantonese words (all beginning with /h/) read out to him as follows: 希 /hei/ → حي /hiː/; 哈 /ha/ → ﺣﺎ /haː/; 圈 /hyn/ → ﻫﻮن /huːn/; 開 /hɔi/ → ﻫﻮىّ /huːj/. The highlighted letters ح (written as ﺣ at the beginning of a word) and ه (written as ﻫ at the beginning of a word) represent /ḥ_/ and /h_/ respectively. It can be seen from here that the Cantonese phoneme /h/ was perceived as two different sounds /ḥ/ and /h/ by our Arabic teacher. The Arabic letters ح and ه can be regarded as the respective counterparts of the Phoenician signs 日 and ∃ : while 日 and ح denote /ḥ_/, ∃ and ه denote /h_/ (see Appendix 2). Just as the Arabic letters ح and ه can be used to write the Cantonese /h_/ syllables, so the Phoenician signs 日 and ∃ could be used to write the Greek /h_/ syllables.

If it is true that the Phoenicians used both 日 and ∃ to write the Greek /h_/ syllables at the outset, why did the Greeks eventually use 日 only to represent their /h_/ syllables? One possible reason is that at a later stage of the development of the Greek alphabetic writing system, the Greeks found it necessary to create a new *mater* for rhyming with syllables ending in /e(ː)/ or /ɛː/. Since the name of the Phoenician sign ∃ was /eː/ (or /ɛː/) in some Greek regions and /heː/ (or /hɛː/) in others (this point will be elaborated in §13), it was a strong candidate for the new *mater* in the alphabet. The Greeks finally chose ∃ as the new *mater* for rhyming with syllables ending in /e(ː)/or /ɛː/. Hence the Greeks used 日 to represent the Greek /h_/ syllables while using ∃ exclusively as a *mater*. The use of ∃ as a *mater*, we believe, is a turning point in the evolution of the Greek writing system. This point will be elaborated on in §12.3.

The representation of the Greek /w_/ syllables

The representation of the Greek /w_/ syllables does not seem to be very complicated. One would expect the Phoenicians to use the sign Y *waw* to write the Greek /w_/ syllables because Y represents the Phoenician /w_/ syllables. But in the local scripts of archaic Greece, the letter that was used to write the Greek /w/ was ٦ / ٦ / ⴼ *wau* / *digamma*, not Υ *waw*. Why? The reason is not difficult to understand if one realizes that ٦ / ٦ / ⴼ is in fact just a variant form of Υ. We believe that initially, Υ was probably used both as a *mater* and as an ordinary or normal sign that represented the /w_/ syllables. Later, when its variant form ٦ / ٦ / ⴼ was used as an ordinary sign that stood for /w_/, Υ could then be used exclusively as a *mater*.

The Phoenician signs eventually used for writing the Greek /k_/, /s_/, /h_/, /w_/ syllables

The Phoenician signs eventually used for writing the Greek /k_/, /s_/, /h_/, /w_/ syllables are listed in Table 9 below. The Greek letters in the third row evolved from the Phoenician signs in the second row.

Table 9 The Phoenician signs eventually used for writing the Greek /k_/, /s_/, /h_/, and /w_/ syllables

Greek syllable	/k_/	/s_/	/h_/	/w_/
Phoenician sign	Ⴤ, φ	Ⴍ	Ⴂ	Υ
Greek letter	Ⴤ/ Κ, Ϙ	Ϩ/ Ϛ, Μ	Ⴂ/Ⴂ / Η	٦ /٦/ ⴼ

The Phoenician signs used for writing Greek CV syllables that had no close counterparts in Phoenician

As was said at the beginning of §11.1.1, many Greek dialects had seventeen consonants, thirteen of which had close counterparts in Phoenician. In other words, archaic Greek had four consonants which did not exist in Phoenician. These four consonants are /ph, th, kh, dz/. How did the Phoenicians write the Greek CV syllables beginning with these consonants? We first deal with the syllables beginning with the aspirated stops /ph, th, kh/.

The Phoenician signs used for writing Greek syllables beginning with /ph, th, kh/

Phoenician had six stops /p, t, k, b, d, g/ while archaic Greek had, besides these six stops, three more, namely /ph, th, kh/. This being the case, the Phoenicians probably heard the Greek syllables beginning with /ph, th, kh/ as sounds closest to their /p_, t_, k_/ syllables. Since the Phoenicians used ⫫, ×, and Ⴤ to represent their /p_, t_, k_/ syllables respectively, one would expect them to use these signs to represent the Greek /pʰ_, tʰ_, kʰ_/ syllables too. As can be seen from the local scripts of archaic Greece, the letters ⌐ / Γ and Ж / K, which evolved from the Phoenician signs ⟩ and ⤬, were used just as expected to represent not only /p/ and /k/ but also /pʰ/ and /kʰ/. However, the letter ⊗ / ⊕ was used to represent /tʰ/ instead of the expected × / Τ. Why did the Phoenicians use the sign ⊗ to write the Greek /tʰ_/ syllables?

Judging from the way in which the /tʰ_/ and /t_/ syllables in foreign names are written in modern Arabic, it was possible for the Phoenicians to use both the signs × and ⊗ to write the Greek /tʰ_/ and /t_/ syllables. The respective Arabic counterparts of the Phoenician signs × and ⊗ are ت and ط. While × and ت denote /t_/, ⊗ and ط denote /tˤ_/ (see Appendix 2). In modern Arabic the /tʰ_/ and /t_/ syllables in foreign names are written generally as ت (read as /t_/) and occasionally as ط (read as /tˤ_/). For example, the /tʰ_/ syllables in the following English names and the /t_/ syllables in the following French names are written in modern Arabic by the letter ت /t_/ (highlighted in the name and written as ﺗ at the beginning of a word): Tony توني, Newton نيوتن; Toussaint توسان, Étienne إتيان. Nevertheless, the /tʰ_/ and /t_/ syllables in some names are written by the letter ط /tˤ_/ (also highlighted in the name): Washington واشنطن; Italia ايطاليا, Αριστοτέλης (Aristoteles) أريسطوطلس, Augustus أغسطس. The /tʰ_/ and /t_/ syllables in foreign names are written in modern Arabic as either ت or ط, possibly because the /tʰ/ or /t/ phoneme in these syllables is perceived as two different sounds by the native speakers of Arabic. Some /tʰ_/ and /t_/ syllables are heard as sounds close to their /t_/ syllables, and so the letter ت is used to denote these sounds; some /tʰ_/ and /t_/ syllables are heard as sounds close to their /tˤ_/ syllables, and so the letter ط is used to denote these sounds.

Our Arabic teacher wrote the Cantonese syllables with the initial consonant /tʰ/ in the following three words as ت /t_/: 湯 /tʰɔŋ/ تونغ, 天 /tʰin/ تن, 太 /tʰai/ تاي. However, he wrote the Cantonese word 灘 /tʰan/ as طان /tˤan/. His perception of the Cantonese word /tʰan/ as a sound close to the Arabic /tˤaːn/ might be due to the fact that in the word /tʰan/, the sound /tʰ/ is affected by the following vowel /a/, which is open and back like the vowel /ɑː/ in the English word *father*.

The use of ت and ط to write the /tʰ_/ and /t_/ syllables in foreign names in modern Arabic indirectly supports the hypothesis that it was possible for the Phoenicians to use both the signs × and ⊗ to write the Greek /tʰ_/ and /t_/ syllables. In Table 10 below are listed the signs that the Phoenicians probably used to write the Greek syllables beginning with the six voiceless stops /pʰ, p, tʰ, t, kʰ, k/:

**Table 10 Phoenician signs used for writing Greek syllables
beginning with /pʰ, p, tʰ, t, kʰ, k/**

Greek syllable	/pʰ_/	/p_/	/tʰ_/	/t_/	/kʰ_/	/k_/
Phoenician sign	⟋	⟋	⊗, ×	⊗, ×	𐤊, 𐤒	𐤊, 𐤒
Phoenician syllable	/p_/	/p_/	/tˤ_/, /t_/	/tˤ_/, /t_/	/k_/, /q_/	/k_/, /q_/

If it is true that the Phoenicians used both the signs × and ⊗ to write the Greek /tʰ_/ and /t_/ syllables, then two problems would arise in the writing and reading of Greek. In writing /tʰ_/ and /t_/, the Greeks would have to decide whether to use × or ⊗. In reading the sign × or ⊗, the Greeks would have to decide whether it represented /tʰ_/ or /t_/. To make life easier, the Greeks probably decided eventually to use only ⊗ to write /tʰ_/ and only × to write /t_/. Consequently, as can be seen from the scripts of archaic Greece, × denoted only /t/, and ⊗ only /tʰ/.

The Phoenician sign used for writing Greek syllables beginning with /dz/

The Phoenician sign 𐤆 *zayin*, we believe, was used to represent the Greek /dz_/ syllables. According to the British linguist W. Sydney Allen, at the time when the Greeks adopted the Semitic alphabet, the sign 𐤆 probably represented the Greek affricate /dz/, which later underwent metathesis in early Greek to become /zd/. The archaic Greek affricate /dz/, Allen argues, probably evolved from a coalescence of /dj/ in Indo-European. How /dj/ could evolve into /zd/ through /dz/ may be illustrated by the possible route of evolution taken by the Indo-European word *pedyos* (1974:53-55):

ped-yos /pedjos/ > archaic Greek */pedʒos/ > */pedzos/ > early Greek πεζός /pezdos/

Thus, according to Allen, the sign 𐤆 was probably read as /dz/ in archaic Greek, and later as /zd/ in early Greek. The coalescence of /d/ and /j/ to become /dʒ/ is a common phenomenon in many languages, and metathesis may arise occasionally in some languages. For example, the English words *waps* and *þrid* (thrid) underwent metathesis to become *wasp* and *third*.

There is no doubt that the sign 𐤆 was read as /zd/ in early Greek because this is supported by inscriptional evidence. One might wonder whether the affricate /dz/ had already evolved into /zd/ at the time when the Greeks adopted the Phoenician alphabet. The Phoenician sign 𐤆, we believe, was used to write the Greek /dz_/ syllables at the time when the Greeks adopted the Phoenician alphabet, not /zd_/. If there had been Greek words or syllables beginning with /zd/ then, the Phoenicians might have used the signs I *zayin* plus ◁ *dāleth* to represent the sound /zd_/, whereas the Greeks might have used the signs ‡ *sāmekh* plus ◁ *dāleth*.[12]

If it is true that word-initial /dz/ had not yet evolved into /zd/ in the Greek language at the time when the Greeks adopted the Phoenician alphabet, which Phoenician sign or signs would be used to represent /dz_/? If the Greeks had had to write /dz_/ on their own after learning the Phoenician alphabet, they might have had difficulty choosing a suitable Phoenician sign to represent syllables beginning with /dz/ because there is no Phoenician sign that represents syllables beginning with this affricate sound. Judging from the fact that the sign I *zayin* was chosen to write the Greek /dz_/ syllables, it was probably the Phoenicians who chose this sign, not the Greeks, for the following reasons.

If the Greeks had had to write their /dz_/ on their own, different Greeks might have chosen different Phoenician signs. If the Phoenicians had to write the Greek /dz_/, probably they would generally use the sign *zayin* I only. Under the influence of their mother tongue, the Phoenicians would probably perceive the Greek sound /dz_/ as a sound closest to the Phoenician /z_/ and so would use the sign I *zayin* to represent /dz_/. The Phoenicians would tend to miss the initial /d/ in their perception of the Greek word-initial /dz/, just as native Arabic speakers today tend to miss the initial /d/ in their perception of the Italian word-initial /dz/. The Arabic speaker's perception of the Italian /dz/ is probably a modern parallel of a Semitic speaker's perception of the Indo-European word-initial /dz/. The Italian name Zola /dzɔːla/, for example, is transliterated today in Arabic as ﻻ‌ﻭ‌ﺯ <a?_l uw_z> (Arabic transliterations to be read from right to left hereafter in accordance with the direction of Arabic writing) /zuːla/. The initial /d/ in /dzɔːla/ is ignored in the Arabic transliteration, which means that the initial /d/ is either missed or considered to be too different from the Arabic /d/ or any other Arabic consonant to be so identified. Allen's hypothesis about the existence of the Greek affricate /dz/ at the time when the Greeks adopted the Phoenician alphabet is probably correct.

Jeffery points out that in Thera the Phoenician sign ‡ was used to represent /dz/ instead of I (1961: 317). The Greek affricate /dz/ spoken in Thera might have been less voiced than in other dialects and so was perceived by the Phoenicians as /s/. Hence the Phoenicians used the sign ‡ meant for writing /s_/ to represent the Theran

---

12 See Allen (1974:54-55) as to why the Greeks might have chosen the signs σδ <sd> to represent /zd/ if /zd/ had existed at the time when the Greeks first adopted the Phoenician alphabet.

/dz_/ syllables. It goes without saying that the Therans would read the sign ‡ as /dz_/.

The Phoenician signs used to write the Greek CV syllables: a sum-up

To write a Greek CV syllable, the Phoenicians would use a Phoenician sign plus a *mater*. Table 11 below shows the eighteen different Phoenician signs that the Phoenicians would use when writing the Greek CV syllables. The highlighted Greek syllable onsets /p^h/, /t^h/, /k^h/, and /dz/ did not exist in Phoenician, and the Phoenician signs ‡, ᵡ, and ∃ representing /s_/ and /h_/ were discarded later on by the Greeks, while ⊗ and × were used to represent /t^h_/ and /t_/ respectively.

Table 11 Signs that the Phoenicians probably used to write the Greek CV syllables

Greek	p^h_,	b_	t^h_,	d_	k^h_,	g_	m_	n_	s_	dz_	l	r	h_	w_
syllable	p_		t_		k_									
Phoenician	⟩	⟨	⊗,	◁	ᵞ,	⟨	ᵐ	⟨	ꟺ, I	⫶	ᘓ	⟩	⊟,	Υ
sign			×		φ					‡,				∃
										ᵡ				

The importance of the Phoenicians' perception of Greek sounds

In regard to the genesis of the Greek alphabet, the Phoenicians' perception of Greek sounds seems to have been overlooked. If one allows for the possibility of the Phoenicians' leading role in writing the Greek sounds at the earliest stage of Greek alphabetic writing, one can account for the choice of the Phoenician signs for writing Greek syllables as listed in Table 11. One can also explain why the local scripts of archaic Greece, if they started with the signs as listed in Table 11 at the earliest stage, could end up with such signs as presented in Jeffery's Table of Letters (1961: at end).

Although the correspondence between the Greek and Phoenician consonants is close, it is not always an easy one-to-one correspondence. A Greek consonant may be heard by the Phoenicians as different sounds corresponding to two or even three Phoenician consonants, and two different Greek consonants may be heard as the same sound corresponding to one Phoenician consonant. Thus the Phoenicians might have used two or three signs to represent Greek syllables beginning with the same consonant, and only one sign to represent Greek syllables with different initial consonants, as can be seen from Table 11 above.

One can perceive from this section (§11.1) that there are three kinds of correspondence between the Greek and Phoenician consonants, which will be shown in Tables 12.1, 12.2 and 12.3 below. The first row of each table shows the Greek syllables to be written by the Phoenicians; the second row shows the Phoenician syllables that correspond to the Greek syllables in the first row; the third row shows the Phoenician signs that the Phoenicians would use to write the corresponding Greek syllables in the first row.

Table 12.1 Cases of a one-to-one correspondence between the Greek and Phoenician consonants

Greek syllable:	/b_/	/d_/	/g_/	/m_/	/n_/	/l_/	/r_/	/w_/	/dz_/
Phoenician syllable:	/b_/	/d_/	/g_/	/m_/	/n_/	/l_/	/r_/	/w_/	/z_/
Phoenician sign:	ᕴ	◿	∧	ᙈ	ካ	ʅ	◁	Υ	ℐ

Table 12.2 Cases of a one-to-two/three correspondence between the Greek and Phoenician consonants

Greek syllable:	/tʰ_/		/t_/		/kʰ_/	/k_/		/s_/			/h_/	
Phoenician syllable:	/t_/	/tˤ_/	/t_/	/tˤ_/	/k_/	/q_/	/k_/	/q_/	/s_/	/ʃ_/	/sˤ_/	/h_/	/ħ_/
Phoenician sign:	×	⊗	×	⊗	⅄	φ	⅄	φ	‡	w	ℽ	∃	⊟

Table 12.3 Cases of a two-to-one correspondence between the Greek and Phoenician consonants

Greek syllable:	/pʰ_/	/p_/	/tʰ_/	/t_/	/tʰ_/	/t_/	/kʰ_/	/k_/	/kʰ_/	/k_/
Phoenician syllable:	/p_/		/t_/		/tˤ_/		/k_/		/q_/	
Phoenician sign:	?		×		⊗		⅄		φ	

11.1.2 The Phoenician way of using *matres* to write Greek CV syllables

As was said earlier, the Phoenicians would use a Phoenician sign plus a *mater* to write a Greek CV syllable. We argued earlier in this book that the Phoenicians had three *matres* ᐖ, ᔭ, and Υ. Which *mater* to use would depend on the V of the Greek

CV syllable to be written.

Ancient Greek has many more vowels than Phoenician. As was said earlier, Phoenician possibly had only eight monophthongs: /a(ː), i(ː), u(ː), eː, oː/. Ancient Greek, however, had not only a lot of monophthongs, but also quite a lot of diphthongs.

Ancient Greek had 10-12 monophthongs, subject to dialect variation. Some dialects had ten monophthongs, which are generally transcribed as: /a(ː), i(ː), u(ː), e(ː), o(ː)/. Five of the monophthongs were short, and five were long. The quality of a short monophthong is presumed to be basically the same as that of a long monophthong. It should be noted that the mid vowels /e(ː), o(ː)/ might be realized as vowels that lay somewhere between [eː, oː] and [ɛː, ɔː]. If /e(ː), o(ː)/ were realized as vowels that were closer to [ɛː, ɔː] than to [eː, oː], then /e(ː), o(ː)/ could well be transcribed as /ɛ(ː), ɔ(ː)/. It is also possible that in some dialects the short mid vowels differed in quality from the long mid vowels. The short and long mid vowels might differ so much in quality that they could be regarded as different vowel qualities. In such cases, the short mid vowels can be transcribed as /e, o/ and the long mid vowels as /ɛː, ɔː/. Some Greek dialects had twelve monophthongs, which are generally transcribed as /a(ː), i(ː), u(ː), e(ː), o(ː), ɛː, ɔː/ (Malikouti-Drachman 2007:526; Teffeteller 2006:150; for the tongue position for the vowels, see Figure 60 in Petrounias 2007:558).

Ancient Greek had 10 diphthongs, which are generally transcribed as: /ei, ɛːi, ai, aːi, ui, oi, ɔːi, eu, au, ou/. Generally speaking, these diphthongs were not used as frequently as the monphthongs.

A Greek CV syllable may consist of a consonant plus a monophthong or a diphthong. To represent this type of syllable, the Phoenicians would use a Phoenician sign plus a *mater*.

Regarding the choice of *matres* in representing a CV syllable, it is the *timbre* or quality of the vowel that would determine which *mater* to use to represent the syllable, not the length of the vowel. Thus, regardless of the length of the V in the CV, the Phoenicians would use the same *mater* as long as the quality of V remained the same.

If the initial C of a Greek CV syllable is /n/, it is possible in theory for it to combine with each of the above monophthongs or diphthongs to form the following twenty-two syllables: /na(ː), ni(ː), nu(ː), ne(ː), no(ː), nɛː, nɔː, nei, nɛːi, nai, naːi, nui, noi, nɔːi, neu, nau, nou/. How would the Phoenicians write such syllables in a Greek name? How would they decide on which *mater* to use after the sign ﬠ /n_/?

As the Phoenicians had only three *matres* in their alphabet, they could write the above twenty-two syllables only in these three written forms: 𐤊𐤍, 𐤆𐤍, and 𐤅𐤍, which they would read basically as /na(:)/, /ni(:)/, and /nu(:)/ respectively. However, as the Phoenicians probably had /e:/ and /o:/ apart from /a(:)/, /i(:)/ and /u(:)/, they would hear the Greek syllables /ne:/ and /ni:/ as two distinct syllables, and so would they hear the Greek syllables /no:/ and /nu:/. If they had to write a foreign speech sound that was close to their /ne:/, they would be obliged to write it as 𐤆𐤍, since the /ni(:)/ sound of 𐤆𐤍 was the closest to this foreign speech sound, not the /nu(:)/ sound of 𐤅𐤍, nor the /na(:)/ sound of 𐤊𐤍. A problem, however, might arise when they read back 𐤆𐤍 sometime later. If they remembered the original sound correctly, they would probably read 𐤆𐤍 as /ne:/. If they could not remember the original sound, they might read it as /ni:/. In other words, the Phoenicians might sometimes read 𐤆𐤍 as /ne:/ apart from /ni(:)/. By the same token, they might sometimes read 𐤅𐤍 as /no:/ apart from /nu(:)/. Thus it is possible that 𐤆𐤍 and 𐤅𐤍 each basically had two readings while 𐤊𐤍 had one reading only.

If the above hypothesis is correct, then in writing a CV syllable beginning with /n/ in a foreign name, the Phoenicians would compare this sound with their /na(:)/, /ni(:)/, /nu(:)/, /ne:/, and /no:/ syllables. Any sound that was close to their /na(:)/ would be written as 𐤊𐤍; any sound that was close to their /ni(:)/ or /ne:/ would be written as 𐤆𐤍, and any sound that was close to their /nu(:)/ or /no:/ would be written as 𐤅𐤍. The Phoenician way of writing foreign speech sounds had probably become a method of sound comparison, which was one step beyond the rhyming principle governing the use of *matres*.

The Phoenician sounds /na(:)/, /ni(:)/, /nu(:)/, /ne:/, and /no:/ were possibly the yardsticks against which the above twenty-two Greek syllables were to be measured when the Phoenicians had to decide which of the above three written forms to use in writing these syllables. They would have no difficulty in writing /na(:)/, /ni(:)/, and /nu(:)/, because the three *matres* rhyme readily with these syllables. As to the remaining syllables, none of the three *matres* rhyme easily with them, but even so they would have to decide which of the three written forms 𐤊𐤍, 𐤆𐤍, and 𐤅𐤍 could best represent these syllables. For example, they might use 𐤆𐤍 for /ne:/ by reasoning like this: since the Phoenician sound /ne:/ of 𐤆𐤍 was the closest to the Greek sound /nɛ:/, they could write /nɛ:/ as 𐤆𐤍. They would probably write the other syllables by using the same method of sound comparison. As there are no surviving written records to show how the Phoenicians wrote a Greek name, we have to resort to a modern Semitic script to explore the possible ways in which the Phoenicians would set about writing it.

The Arabic script is a good choice for this purpose as Phoenician and Arabic are both Semitic languages, and as the Phoenician writing system probably has three *matres*, just like Arabic. The native Arabic speakers' perception of the sounds of a foreign name and their choice of *matres* might shed some light on the possible way in which the Phoenicians perceived and wrote a Greek sound. It must be stressed here that the Phoenician vowel system is not the same as that of modern Arabic: while Phoenician possibly has eight monophthongs /a(:), i(:), u(:), e:, o:/, Classical Arabic has only six monophthongs /a(:), i(:), u(:)/ and two diphthongs /ai, au/. The Phoenician consonant system is not the same as that of modern Arabic either, despite many similarities. Thus the Phoenicians' perception of foreign speech sounds may not be entirely the same as that of the Arabic speakers. Even so, the Arabic speakers' perception of similar sounds in foreign languages is a useful source of reference in estimating how the Phoenicians would write the above-mentioned twenty-two Greek CV syllables. We may be able to infer the Phoenician way of writing foreign speech sounds by trying to see how these sounds are actually written in Arabic. We will deal with the Phoenician way of writing those Greek syllables containing monophthongs first and then those containing diphthongs later, as the latter case is more complicated.

Table 13 below shows how the Phoenicians would write a Greek CV syllable in which V is a monophthong. Based on how similar syllables in a foreign name are written in Arabic, we assume that eventually the Phoenicians would write these syllables with the three *matres* as shown in the second column of the table. As can be seen from the table below, the Arabic letter ن /n_/ (written as نـ at the beginning of a word) is the counterpart of the Phoenician sign 케 /n_/. The Arabic letters ا, ء and و are the respective counterparts of the Phoenician *matres* ⱪ, ⱦ and Ⲩ. Please be reminded that Arabic transliterations are to be read from right to left in accordance with the direction of Arabic writing.

Table 13 The Phoenician way to write a Greek CV syllable
(V = a monophthong)

Greek syllable	Phoenician		Probable modern parallel: Arabic				
	Written form	Sound	Written form	Sound	Foreign name	Foreign name in Arabic	Transliteration of foreign name written in Arabic with the relevant *mater* highlighted
/na(:)/	✦ꟻ	/na(:)/	نا	/na(:)/	Nasdaq (eng)	ناسداك	< _k *aʔ*_d _s *aʔ*_n>
/ni(:)/	⁀ꟻ	/ni(:)/	نِ	/ni(:)/	Nice (fre)	نيس	< _s *ij*_n>
/nu(:)/	ⵝꟻ	/nu(:)/	نو	/nu(:)/	Louvre (fre)	لوفر	< _r _f *uw*_l>
/ne(:)/	⁀ꟻ	/ne:/	نِ	/ni(:)/	Bell (eng)	بيل	< _l *ij*_b>
/no(:)/	ⵝꟻ	/no:/	نو	/nu(:)/	Beaumarchais (fre)	بومارشيه	< _h *ij* ʃ *aʔ*_m *uw*_b>
/nɛ:/	⁀ꟻ	/ne:/	نِ	/ni(:)/	Blaise (fre)	بليز	< _z *ij*_l _b>
/nɔ:/	ⵝꟻ	/no:/	نو	/nu(:)/	Ross (eng)	روس	< _s *uw*_r>

As can be seen from the above table, the Phoenicians might eventually use ✦ꟻ for the Greek syllable /na(:)/, ⁀ꟻ for /ni(:), ne(:), nɛ:/, and ⵝꟻ for /nu(:), no(:), nɔ:/. The Phoenicians might read ⁀ꟻ as /ni:/ or /ne:/, and ⵝꟻ as /nu:/ or /no:/. However, as Classical Arabic, like ancient Egyptian, basically has only three vowels (discounting vowel length): /a, i, u/, نِ and نو, the counterparts of ⁀ꟻ and ⵝꟻ, have basically only one reading each. They are read as /ni(:)/ and /nu(:)/ respectively.

As foreign names containing the syllables /nu(:), ne(:), no(:), nɛ:, nɔ:/ are hard to find, we are obliged to use as examples those foreign names containing syllables with the same rhymes. Even so, we can see from these examples which *matres* modern Arabic would use if it had to write the syllables /nu(:), ne(:), no(:), nɛ:, nɔ:/ in foreign names.

It should be noted that in the table above the sounds /lɛ/ and /rɔ/ in the names *Blaise* and *Ross* are written in Arabic as لي and رو, which are transliterated as <ij_l> and <uw_r> in this book. The written forms لي and رو here are read in Arabic as /li:/ and /ru:/, which are quite different from the original sounds /lɛ/ and /rɔ/.

We now deal with those Greek syllables containing a diphthong. In the left-hand column of Table 14 below are listed ten Greek syllables each containing a diphthong. In writing out these ten syllables, the Phoenicians would probably compare these sounds with the five Phoenician syllables having the same onset: /naː, niː, nuː, neː, noː/, just in the same manner as they would write those Greek syllables containing a monophthong. How the Phoenicians might write these ten syllables and read their written forms is shown in the second column of Table 14. A question mark (?) next to the sound means that it is uncertain how the Phoenicians would read the written form.

Table 14 The Phoenician way to write a Greek CV syllable (V = a diphthong)

Greek syllable	Phoenician		Probable modern parallel: Arabic				
	Written form	Sound	Written form	Sound	Foreign name	Foreign name in Arabic	Transliteration of foreign name written in Arabic with the relevant *mater* highlighted
/nei/	ר	/ni:/?	نِي	/ni:/	Bailey (eng)	بيلي	<ij̲ l *ij̲* b>
/nou/	ר	/nu:/?	نو	/nu:/	Foley (eng)	فولي	<ij̲ l *uw* f>
/na(:)i/	ר	/na:/	نايِ	/na:j/	Leipzig (ger)	لايبزيغ	<γ *ij̲* z b j aʔ l>
/nau/	ר	/na:/	ناو	/na:w/	Braun (ger)	براون	<n waʔ r b>
/neu/	ר	/ni:/?	نيو	/ni:w/	Ζεύς (gre)	زيوس	<s w*ij̲* z>
/nɛːi/	ר	/neː/	(A modern European name containing the diphthong /ɛːi/ is hard to find. As /nɛːi/ is fairly close to /nei/, the written form of /nei/ may also be appropriate for writing /nɛːi/.)				
/noi/	ר	/nu:/?	(A modern European name containing the diphthong /oi/ is hard to find. As /noi/ is fairly close to /nɔːi/, the written form of /nɔːi/ may also be appropriate for writing /noi/.)				
/nɔːi/	ר	/no:/	نو ي	/nu:j/	Illinois (eng)	إلينوي	<*juw* n *ij̲* l iʔ>
/nui/	ר	/nu:/	نو ي	/nu:j/	Louis (fre)	لوي	<*juw* l>

The Phoenicians would probably not be able to analyse the sound of a diphthong in the same manner as modern Arabic does. This should not be very surprising as the Arabic script has benefited from the knowledge of sound analysis accumulated during its long history of development. The Phoenicians would probably be obliged to use the method of sound comparison in writing out a Greek syllable containing a diphthong. They would try to find out which of the five relevant Phoenician syllables ending in /_aː, _iː, _uː, _eː, _oː/ was aurally the closest to this syllable. Take the Greek syllable /nei/ for example. They would compare /nei/ with the five Phoenician syllables /na:, ni:, nu:, ne:, no:/ in order to decide which of the five Phoenician syllables was the closest to /nei/. To the Phoenicians, the Greek /nei/ might sound the closest to either their /ni:/ or /ne:/, depending on the actual sound value of the Phoenician /e:/. If the Phoenician /e:/ sounds like the first vowel of the French word *été*, the Phoenicians might hear the Greek /nei/ as a sound that their /ne:/ was the closest to. They might then write it as ר and read ר as /ne:/. On the other hand, if the Phoenician /e:/ sounds like the first vowel of the French word *être*, the Phoenicians might hear the Greek /nei/ as a sound that their /ni:/ was the closest to. They might then write it as ר and read ר as /ni:/. As was said earlier, how the Phoenicians might write a Greek syllable containing a diphthong and read the written form is shown in the second column of Table 14. As the actual sound values of the Phoenician /e:/ and

/oː/ are uncertain, we are not quite sure how the Phoenicians would read the written forms 𐤍𐤉 and 𐤍𐤅 in the second column of Table 14. As can be seen from Table 14, the Phoenician way of writing a syllable containing a diphthong is generally quite different from the Arabic one.

11.2 The Phoenician way to write a Greek CVC syllable

How would the Phoenicians write a Greek name with a CVC syllable structure, such as /nestɔːr/? They might write a CVC syllable initially as: Sign_1 + $\textit{Mater}_1$ + Sign_2 + $\textit{Mater}_2$, thus turning the CVC structure into CVCV. They might write the first CVC syllable /nes/ as 𐤍𐤉𐤎𐤉, and read it as /neːsi/ or /niːsi/. The Phoenicians might write the /s/ sound as 𐤎𐤉, because they might hear the Greek /s/ as a sound that the /si(ː)/ sound of 𐤎𐤉 was the closest to. On the other hand, the Phoenicians might sooner or later come to realize that, if they dropped the last *mater*, the CVC syllable would be more accurately represented, because in Phoenician writing the sign 𐤎 sometimes stood for /s/ followed by a barely perceptible schwa. In other words, when not followed by a *mater*, 𐤎 would be read as /s/ followed by a barely perceptible schwa. Likewise, they might write the second CVC syllable /tɔːr/ initially as 𐤕𐤅𐤓𐤀 /toːra/ or /tuːra/, and eventually as 𐤕𐤅𐤓 /toːr/ or /tuːr/. To the Phoenicians, the norm for writing a Greek CVC syllable would ultimately be: Sign_1 + $\textit{Mater}_1$ + Sign_2. Sign_2 could be read as a light and short sound, albeit with different degrees of accentuation. This sound, if analysed phonemically, can be regarded as a C followed by a barely audible schwa or no schwa. It goes without saying that the way to read Sign_2 should conform to the Phoenician phonological system.

11.3 The Phoenician way to write a Greek V syllable

Judging from the spelling of ancient Greek names in the local scripts of archaic Greece, many of such names began with a V syllable. This word-initial syllable consisted of only one vowel, which was either a monophthong or a diphthong. The word-initial V syllables commonly used in ancient Greek names are: /a(ː), i(ː), u(ː), e(ː), o(ː), ɛː, ɔː, ai, au, ui, ei, eu, oi, ou/. How would the Phoenicians write these syllables, when they did not have any words beginning with V syllables in their language? By examining the way in which foreign names with word-initial V syllables are represented in the modern Arabic script, one can hazard a guess at the possible ways in which the Phoenicians tackled the task.

As can be seen in Table 15 below, a word-initial V syllable in a foreign name is treated as a CV syllable comprising a glottal stop /ʔ/ + a V in the Arabic script. This being the case, it is possible that under the influence of their mother tongue, the Phoenicians would hear a Greek V syllable as a CV syllable comprising a glottal stop /ʔ/ + a V. For example, the Phoenicians would probably perceive the word-initial syllables /a/, /i/, and /u/ in Greek as sounds close to their /ʔa/, /ʔi/, and /ʔu/ respectively, and so would write the Greek V syllables /a/, /i/, and /u/ as ⟨ᚴᚴ⟩, ⟨ᚴ⟩, and ⟨Υᚴ⟩ accordingly. The first sign of each of the above three written forms is ᚴ, which stands for the /ʔ_/ syllables. The second sign is the *mater* that specifies the sound value of ᚴ /ʔ_/. Table 15 below shows how the Phoenicians might write the Greek word-initial V syllables.

To write the Greek word-initial V syllables, the Phoenicians might use ᚴᚴ /ʔa(:)/ for /a(:), ai, au/, ᚴ /ʔi(:)/ or /ʔe:/ for /i(:), e(:), ɛ:, ei, eu/, and Υᚴ /ʔu(:)/ or /ʔo:/ for /u(:), o(:), ɔ:, ou, oi, ui/.

It should be noted that the Arabic way of writing a foreign V syllable is a very much developed one. As can be seen from Table 15, the ordinary or normal sign that represents the word-initial /ʔ_/ syllable in a foreign name seems to have been differentiated by the use of diacritics. The sign ا very often stands for /ʔa/, but it may also stand for /ʔu/ or /ʔ_/. Its sound value is to be determined by the reader. The signs إ and أ, however, are less ambiguous: they stand for /ʔi/ and /ʔa:/ respectively. The above signs can be followed by a *mater*. For example, إي and أو stand for /ʔi:/ and /ʔu:/ respectively. It can be said that when compared with the Arabic way of writing a V syllable in a foreign name, the Phoenician way of writing a Greek V syllable is perhaps generally more primitive or primary.

Table 15 The Phoenician way to write a Greek V syllable

Greek syllable	Phoenician		Probable modern parallel: Arabic				
	Written form	Sound	Written form	Sound	Foreign name	Foreign name in Arabic	Transliteration of foreign name written in Arabic with the relevant *mater* highlighted
/a(:)/	𐤊𐤊	/ʔa(:)/	ا	/ʔa/	Apollo (gre)	أبولو	<*uw*_l *uw*_b a?>
/i(:)/	𐤉𐤊	/ʔi(:)/	إِﻳ	/ʔiː/	Icarus (gre)	إيكاروس	<_s *uw*_r a?_k *ij*_?>
/u(:)/	𐤉𐤊	/ʔu(:)/	أو	/ʔuː/	Ulysses (lat)	أوليسيس	<_s *ij*_s *ij*_l *uw*_?>
/e(:)/	𐤉𐤊	/ʔe:/	اِ	/ʔi/	Émile (fre)	إميل	<_l *ij*_m i?>
/o(:)/	𐤉𐤊	/ʔo:/	أو	/ʔuː/	Olympia (gre)	أولمبيا	<_? *ij*_b_m_l *uw*_?>
/ɛ:/	𐤉𐤊	/ʔe:/	إِﻳ	/ʔiː/	Aerosmith (eng)	إيروسميث	<_θ *ij*_m_s *uw*_r *ij*_?>
/ɔ:/	𐤉𐤊	/ʔo:/	أو	/ʔuː/	Austin (eng)	أوستن	<_n_t_s *uw*_?>
/ai/	𐤊𐤊	/ʔa:/	أَﻳ	/ʔai/	Ireland (eng)	أيرلندا	<a?_d_n_l_r_ja?>
/au/	𐤊𐤊	/ʔa:/	أو	/ʔu/	Augustus (lat)	أغسطس	<_s_tˤ_s_y u?>
/ei/	𐤉𐤊	/ʔi:/?	أَﻳ	/ʔa:j/	Acer (eng)	آيسر	<_r_s_ja?>
/eu/	𐤉𐤊	/ʔi:/?	أو	/ʔu/	Europa (lat)	أوروبا	<a?_b *uw*_r *uw*_?>
/ou/	𐤉𐤊	/ʔu:/?	أو	/ʔu:/	O'Connor (eng)	أوكونور	<_r *uw*_n *uw*_k *uw*_?>
/oi/	𐤉𐤊	/ʔo:/?	(A modern European name containing the word-initial diphthong /oi/ is hard for us to find in modern Arabic.)				
/ui/	𐤉𐤊	/ʔu:/	(A modern European name containing the word-initial diphthong /ui/ is hard for us to find in modern Arabic.)				

11.4 The Phoenician way to write a Greek VC syllable

How would the Phoenicians write the Greek VC syllables? The Phoenicians might turn the VC syllables into CVCV or CVC. The possible ways in which the Phoenicians wrote CVCV or CVC syllables have been explained above.

11.5 The Phoenician way to write a Greek consonant cluster

The Greek language has consonant clusters, which the Semitic languages generally lack. How would the Phoenicians write the consonant clusters in such a Greek name as /kle-o-pa-tra/? Some Phoenicians might turn a C_1C_2 cluster into a single C by leaving out the less prominent C. The syllable /kle/ might be written as 𐤉𐤘 /ki:/ or /ke:/, and the syllable /tra/ as 𐤊𐤗 /ta:/. Phoenicians who had sharp ears might write the first syllable /kle/ as 𐤉𐤋𐤘 /kili:/ or /kile:/ and the last syllable /tra:/ as 𐤊𐤓𐤊𐤗 /tara:/, by turning each CCV syllable into two CV syllables. They might realize at a later stage that they could write the Greek sounds /kle/ and /tra/

more accurately by leaving out the first *mater* of each written form. They would read ꓫꛯꚛ and ꓘꛩ× as /kəli:/ (or /kəle:/) and /təra:/ respectively. When reading ꓫꛯꚛ and ꓘꛩ×, they would try to pronounce the schwa /ə/ as lightly as possible.

11.6 How the Phoenicians would write and read a Greek name

Based on their ways of writing various syllable structures in Greek names that have been discussed so far, the Phoenicians might write and read the following four Greek names as shown in Table 16.

Table 16 The Phoenician ways to write and read Greek names

Greek name	/nika:/	/nesto:r/	/kleopatra/	/aristotelɛ:s/
Written as	ꓘꚛꛯꓩ	ꛩꙄ×ꜟꛯꓩ	ꓘꛩ×ꓘꝘꚛꚛꛯꚛ	ꜟꛯꙆ×ꚛ×ꜟꛯꛩꓘꓘ
Read as	/nika:/	/ni:sətu:rə/ or /ne:səto:rə/	/kəli:wpa:təra:/ or /kəle:wpa:təra:/	/ʔarisətutile:s/ or /ʔarisəto:tile:s/

The Phoenician ways to write and read Greek names would be conditioned by both their method of writing foreign names and their perception of Greek sounds. The closeness of the Phoenicians' reading of a written Greek name to the actual Greek pronunciation should be of secondary importance as long as the Phoenicians could identify the person or place that the name stood for when reading out the written name.

12 From the Phoenician way of writing Greek names to the Greek way of writing the Greek language

12.1 From the writing of Greek names to proto-Greek alphabetic writing

As was said earlier, the Phoenicians might have needed to record Greek names in book-keeping in the second millennium BC. In theory they could write a Greek name either with or without *matres*. But in practice the Phoenicians would soon come to realize that a Greek name written with *matres*, being less ambiguous in sound representation, was much easier to read out or read back afterwards than one written without *matres*. Thus the Phoenicians would probably decide at an early stage to use *matres* to write Greek names. However, their way of writing Greek names might have escaped the notice of the Greeks at the earliest stage because the Greeks had not yet realized that it could be put to some significant use in their daily lives.

Towards the end of the second millennium BC, because of the intense contacts between the Phoenicians and the Greeks, some Greeks might come to realize that the Phoenicians only used about twenty signs to write Greek names in book-keeping. They would be amazed to find that any Greek names could easily be written by means of the Phoenician signs and that some Greek names, when read out by the Phoenicians, sounded pretty close to their own pronunciations. As regards Greek names, it should be noted that a Greek name might comprise everyday Greek words. For example, the Greek names *Aristoteles* and *Cleopatra* mean respectively 'the best purpose' and 'glory of the father'. These names are short phrases that involve the use of Greek syntax. So sooner or later some Greeks were going to realize that they could expand these phrase-like names into short sentences and write out the entire Greek language by following the Phoenician way of writing Greek names with only twenty-odd Phoenician signs. This idea might have motivated the Greeks to learn writing, possibly in the tenth century BC, from the Phoenicians.

To use a foreign alphabet to write one's native language, it is easier for one to learn from native users of that alphabet how they use their alphabet to write out the words of the language than for one to try to write the language on one's own after learning the alphabet. Similarly, it would be easier for the Greeks to learn direct from the Phoenicians how to write the sounds of Greek in the Phoenician alphabet than for the Greeks to try to write Greek on their own after learning the Phoenician alphabet. Many people assume that the Greeks attempted to write their language on their own after learning the Phoenician alphabet. However, judging from the Phoenician signs that were adopted to write archaic Greek, it seems that the Greeks simply followed the Phoenicians' lead in writing their own language at the earliest stage, as has just been explained in §11.

To learn to write Greek in Phoenician signs, some Greeks might ask a Phoenician how Greek names were written and why they were so written. The Phoenician might write out a Greek name, say, /nika:/ as ꓘꓵꓕꓶ and then explain how the signs functioned in the written name. We think that the Phoenician might explain in the following way.

The first sign ꓶ, called *nūn* (or the like), stood for several sounds in Phoenician, which he would probably say aloud for the Greeks: /na(:)/, /ni(:)/, /nu(:)/, /ne:/, /no:/. All the Greek "sounds" that sounded like the above Phoenician sounds would be written as ꓶ.

The second sign ꓕ, called *yōd* (or the like), was a rhyme indicator (technically called a *mater* in this book) that required the preceding sign ꓶ to rhyme with it. ꓕ, read as /ji/ in its role as a rhyme indicator here, required ꓶ to be read as /ni(:)/. Therefore, ꓕꓶ would be read as /ni(:)/.

The third and fourth signs ꓘꓵ functioned in the same way as the first and second signs ꓕꓶ. The third sign ꓵ stood for /ka(:)/, /ki(:)/, /ku(:)/, /ke:/, and /ko:/, and the fourth sign ꓘ required ꓵ to rhyme with it. ꓘ, read as /ʔa/, required ꓵ to be read as /ka(:)/. Therefore, ꓘꓵ would be read as /ka(:)/. The whole name ꓘꓵꓕꓶ should be read as /nika:/.

The writing of the Greek name /nika:/ as ꓘꓵꓕꓶ involves the use of two principles. The first is the principle of the multiple sound values of a Phoenician sign; the second is the rhyming principle of a *mater*. These two principles should not be too difficult for the Greeks to understand. Once the Greeks understood why /nika:/ was written as ꓘꓵꓕꓶ, they had already learned the two most fundamental principles of proto-Greek alphabetic writing.

If the Greeks could further understand why the name /nesto:r/ could be written as ꓩꓬꓫꓱꓕꓶ, they would be able to grasp all the basic principles of proto-Greek alphabetic writing. Regarding the written form ꓩꓬꓫꓱꓕꓶ of the name /nesto:r/, the Phoenician might explain that the signs 𐤍𐤉 and 𐤕𐤅 would be read as /ni(:)/ and /tu(:)/ according to the rhyming principle. However, they could also be read as /ne:/ and /to:/. The sound /ne:/ in /nesto:r/ would be written as 𐤍𐤉, but not as 𐤊𐤉 or 𐤕𐤉, because among the sounds of 𐤊𐤉, 𐤍𐤉 and 𐤕𐤉 (i.e. /na(:), ni(:), nu(:)/), the sound /ni(:)/ was the closest to /ne:/. By the same token, the sound /to:/ in /nesto:r/ would be written as 𐤕𐤅. Thus 𐤍𐤉 could be read as /ni(:)/ and /ne:/, and 𐤕𐤅 as /tu(:)/ and /to:/. That 𐤍𐤉 and 𐤕𐤅 each had two basic readings could be regarded as an extension of the rhyming principle.

The signs 𐤎 and 𐤓 in the name 𐤓𐤕𐤎𐤍𐤉 are not followed by any *matres*. In this case, they should be read as /sə/ and /rə/. The strength of the schwa /ə/ could vary, but /sə/ and /rə/ should be spoken in such a way that they sounded like the original sounds of the name. This is the principle of the sound of a Phoenician sign not followed by a *mater*.

The Greeks, through the observation of a lot of Greek names written in Phoenician and through their attempts to write various Greek names in Phoenician signs, would gradually come to grasp how the Phoenician alphabet worked.

As regards the reading of the Greek names written in the Phoenician alphabet, it must be noted that the Greeks would not follow the Phoenicians' way of reading their names, which was bound to be different from the native speakers' pronunciations of the names to a greater or less extent owing to the difference between the Phoenician and Greek phonological systems and to the limitation of the Phoenician writing system in representing foreign sounds. A Greek name read in any way other than the Greek way of pronouncing it would sound foreign or unnatural to the Greek ear. Thus the Greeks would naturally read their names in their own native way. This, however, would bring about a new relationship between the Phoenician written forms and their corresponding Greek sound values. For example, while the Phoenicians would read the sign 𐤆 as /z_/, the Greeks would read it as /dz_/.

After grasping the basic principles of proto-Greek alphabetic writing, the Greeks would try to write Greek for some practical purposes such as book-keeping. They would find that the sound of a word was fairly accurately represented with only about twenty Phoenician signs. They might then try to write short messages and simple records of events with these signs. As the Greek words were written with *matres*, the three Phoenician *matres* would be used persistently and extensively, and an embryonic form of Greek alphabetic writing would emerge from this. We call this stage of Greek writing with three *matres* proto-Greek alphabetic writing. The idea that it was possible to use only about twenty signs to write the Greek language might have spread from one Greek region to another during the tenth century BC.

12.2 The inadequate number of *matres* in proto-Greek alphabetic writing

There is little doubt that proto-Greek alphabetic writing had various flaws, such as the use of two or three different signs to represent syllables with the same initial consonant (see Table 12.2) and the use of the same sign to represent syllables with two different initial consonants (see Table 12.3). But the flaw that troubled the Greeks most would be the inadequate number of *matres* in their writing. To the Phoenicians, a sign plus a *mater*, such as ✯٦, ১٦, or Υ٦, would have only one or two readings, but the Greeks would find that a sign plus a *mater* could stand for five to nine syllables, as can be shown in Table 17 below.

Table 17 The possible Greek sound values of ✯٦, ১٦ and Υ٦

Phoenician written form	The Phoenicians' way to read it	Possible sound values in Greek
✯٦	/na(ː)/	/na(ː)/, /na(ː)i/, /nau/
১٦	/ni(ː)/ or /neː/	/ni(ː)/, /ne(ː)/, /nɛː/, /nei/, /neu/, /nɛːi/
Υ٦	/nu(ː)/ or /noː/	/nu(ː)/, /no(ː)/, /nɔː/, /nou/, /noi/, /nɔːi/, /nui/

It can be seen from the table above that a Phoenician sign plus a *mater* represented a fairly large number of Greek syllables. If *matres* were not used to write the Greek language, a Phoenician sign could represent as many as twenty-two Greek syllables. Obviously, a name or word written with *matres* would be easier to read than one written without. The Greeks might have known that Phoenician was normally written without *matres* and might have attempted to follow suit. If they ever did so, they would soon realize that it would not be easy to read back what had been written and that it would be better to write with *matres* than to write without. From their experience of reading out written words, the Greeks would feel that it would be even better to write with more *matres*.

The Greek scholar E. Voutiras remarks that the ancient Greeks almost always read aloud (2007:275). The ancient Greeks had a habit of reading aloud probably because words in proto-Greek alphabetic writing were not always easy to identify. Word identification in proto-Greek alphabetic writing was difficult for two reasons. First, words were normally written together without any word spaces or dividers.

Second, at the initial stage of proto-Greek alphabetic writing when all written words were new, it would take the Greeks a lot of time to familiarize themselves with the shapes of their written forms. One way to help identify the written words would be to read them out aloud to see which of the various possible readings would fit in with a particular context. It would be very natural for the Greeks to exploit their mother tongue to facilitate word identification. When reading a written word aloud, the ancient Greeks would treat a Phoenician sign plus a *mater* as a unit. The fewer sound values this unit had, the more easily they could work out its precise sound value. Since proto-Greek alphabetic writing had only three *matres*, the ancient Greeks would at times run into words that they could not easily identify. Despite their attempts to read out these words, they might not be able to get the right sounds of the words that would fit the context. They might wonder then whether the use of more *matres*, in addition to the three existing ones, would make the reading of such words easier.

It can be seen from Table 17 above that the use of three *matres* would not be sufficient to differentiate amongst /ni(:)/, /ne(:)/, and /nɛː/, or amongst /nu(:)/, /no(:)/, and /nɔː/, as 𐤍𐤉 represents /ni(:)/, /ne(:)/, and /nɛː/, and 𐤍𐤅 /nu(:)/, /no(:)/, and /nɔː/. The syllables ending in the mid vowels /e(:)/, /ɛː/, /o(:)/ and /ɔː/ were amongst the most frequently used Greek syllables in the spoken language, but there were no specific *matres* to represent them (see Appendix 9). To represent these syllables, the Greeks had to use the *matres* 𐤉 and 𐤅 that were originally used to represent the sounds ending in /i(:)/ and /u(:)/. For example, the Greeks could only write /ne(:)/ or /nɛː/ as 𐤍𐤉, which was originally the written form for /ni(:)/, and similarly they could only write /no(:)/ or /nɔː/ as 𐤍𐤅, which was originally the written form for /nu(:)/. The Greeks would probably think it better to find two more *matres* so that the frequently-used syllables that ended in /e(:)/ (or /ɛː/) and /o(:)/ (or /ɔː/) could have their specific *mater* to indicate their precise sound value.

The Greeks probably knew that the use of 𐤍𐤉 to represent /ni(:)/ was a true application of the rhyming principle, whereas the use of 𐤍𐤉 to represent /ne(:)/ (or /nɛː/) was made possible only by an extension of this principle. They might wonder whether /ne(:)/ (or /nɛː/) could be better represented by a new *mater* that rhymed precisely with it. The same can be said regarding the use of 𐤍𐤅 to represent /nu(:)/, /no(:)/, and /nɔː/. The Greeks might also wonder whether the syllable /no(:)/ (or /nɔː/) could be better represented by a precise *mater*. The mid vowels /e(:), ɛː, o(:), ɔː/ being amongst the most frequently used vowels in ancient Greek, the Greeks would instinctively feel that by creating two more *matres* for rhyming with syllables ending in the mid vowels, they should be able to read Greek more easily.

The Greeks would then search hard for two more *matres* which they felt were essential to a more faithful representation of their spoken language. Obviously, the sounds of these two new *matres* had to rhyme with /ne(:)/ (or /nɛ:/) and /no(:)/ (or /nɔ:/) respectively. Which two signs of the Phoenician alphabet could satisfy this requirement? How could these two Phoenician signs have the sounds that rhymed with /ne(:)/ (or /nɛ:/) and /no(:)/ (or /nɔ:/)?

From the vowel letters that were used in archaic Greek inscriptions in the second quarter of the first millennium BC, one can infer that the Phoenician signs ⨅ *hē* and O *'ayin* were probably chosen as the fourth and fifth *matres*, possibly in the ninth century BC when the need to create two more *matres* was more and more keenly felt as the use of proto-Greek alphabetic writing expanded. Table 18 below shows the vowel letters that were used in the local scripts of archaic Greece.

Table 18 The vowel letters used in the local scripts of archaic Greece

Vowel letters:	A, ᘔ, Y, ⨅, O	A, ᘔ, Y, ⨅, O, 8 / 8	A, ᘔ, Y, ⨅, O, ⊟	A, ᘔ, Y, ⨅, O, ⊟, Ω
Greek regions which used the vowel letters listed in the first row:	**Central Greece:** Attica, Euboia, Boiotia, Thessaly, Phokis, Lokris, Ozolian and Opoutian, Aigina **Peloponnese:** Argos, Eastern Argolid, Lakonia, Arkadia **The Ionic islands:** Ithake, Kephallenia	**Peloponnese:** Corinth, Megara, Sikyon, Phleious, Kleonai, Tiryns	**The Doric islands** (Southern Aegean): Crete, Thera, etc.	**The Ionic Dodekapolis:** Samos, Miletus, etc. **The Aegean islands (Ionic):** Naxos, Paros, Thasos, etc.

About the vowel letters that were used in the local scripts of archaic Greece, we would like to make the following points:

(1) As far as the representation of vowels is concerned, the archaic Greek alphabets are structurally unified, although the letters of these alphabets may assume quite different shapes. All these alphabets have in common the following five basic vowel monographs: A, ᘔ, Y, ⨅, O (or their variant letter shapes).

(2) The five basic vowel monographs used in archaic Greek inscriptions probably point to one single origin of the archaic Greek alphabets.

(3) We believe that A, ᘔ, Y, ⨅, O had all been used as *matres* before they evolved into vowel letters. A, ᘔ, and Y are traditional *matres* inherited from the Phoenicians; ⨅ and O are the fourth and fifth *matres* created by the Greeks. Why ⨅ and O were chosen as *matres* and how A, ᘔ, Y, ⨅, and O changed from *matres* into vowel letters will be explained in §12.3 and §12.5.

(4) The letter 𝟈 that was used in Corinth and Megara is possibly a variant form of ⊟ or ⅂, and the letter 𝗫 that was used in the neighbouring city Sykion is probably a variant form of the Corinthian 𝟈 . These two points will later be elaborated on.

(5) The vowel letter ⊟ was probably created, possibly in the eighth century BC, in a Greek region where /h/ had been lost. The name of the letter ⊟ was /ɛːta/ there, not /hɛːta/. When Greek alphabetic writing became segmental, a letter would come to represent the initial segment of its name. For example, the letter Λ came to represent the initial segment /a(ː)/ of its name *alpha*, and the letter 𝟈 the initial segment /b/ of its name *bēta*. Hence in regions where the letter ⊟ was called /ɛːta/, ⊟ could be used to represent the initial segment /ɛː/ of its name.

(6) The vowel letter Ω was created, possibly in the eighth century BC, in order to denote /ɔː/. Before the creation of Ω, the letter O was used to represent both /o(ː)/ and /ɔː/. After Ω had been created to represent /ɔː/, O came to represent /o(ː)/ only. The letter Ω was created probably by modifying the shape of the letter O.

12.3 The creation of two more *matres* in early Greek alphabetic writing

Why did the Greeks choose ⅂ and O as *matres*? Some Greeks might have had the idea of using those Phoenician signs whose names could rhyme with /ne(ː)/ (or /nɛː/) and /no(ː)/ (or /nɔː/) as the new *matres*. The Greeks who thought in this way would be able to find only two Phoenician signs that could rhyme with /ne(ː)/ (or /nɛː/), namely ⅂ *hē* and 𝟊 *pē*, but no sign that could rhyme with /no(ː)/ (or /nɔː/). They had to find another way to create a new *mater* for rhyming with /no(ː)/ (or /nɔː/).

One probable reason why the Greeks chose ⅂ *hē* rather than 𝟊 *pē* as the *mater* for rhyming with /ne(ː)/ (or /nɛː/) was that ⅂ *hē* could be used exclusively as a *mater*, but not 𝟊 *pē*. As was said in §11.1.1, it is possible that both ⅂ *hē* and ⊟ *ḥēth* were used to represent the Greek /h_/ syllables. When ⊟ *ḥēth* was used to represent the Greek /h_/ syllables, ⅂ *hē* could then be spared from having to represent /h_/ and thus be used exclusively as a *mater* for rhyming with /ne(ː)/ (or /nɛː/). The advantage of using a sign exclusively as a *mater* in archaic Greek alphabetic writing is obvious because the use of such a sign would facilitate ease of reading. As for 𝟊 *pē*, it could not be used exclusively as a *mater* for rhyming with /ne(ː)/ (or /nɛː/), because there was no substitute sign that could relieve it from having to represent the Greek /p_/ and /pʰ_/ syllables.

The idea of using ⋽ *hē* as the *mater* for rhyming with syllables ending in /e(:)/ (or /ɛː/) seems to have been well received in the Greek world. We believe that the Phoenician sign ⋽, read as /eː/ or /ɛː/ in some Greek regions and as /heː/ or /hɛː/ in others owing to dialect variation, was used at the early stage of Greek alphabetic writing as a *mater*. ⋽ꓶ, for example, would probably be read as /ne(:)/ or /nɛː/, because the *mater* ⋽ /(h)eː/ (or /(h)ɛː/) required ꓶ /n_/ to rhyme with it.

In regions where /h_/ syllables were not spoken, ⋽, read as /eː/ or /ɛː/, could undoubtedly be used exclusively as a *mater* since ⋽ would not be required to write the /h_/ syllables. Even in regions where /h_/ syllables were spoken, ⋽ could still be used exclusively as a *mater* if a substitute sign for writing /h_/ syllables could be found. Presumably, the Greeks could also use the Phoenician ⊟ *ḥēth* to represent /h_/ syllables (see §11.1.1). We believe that in early Greek alphabetic writing, ⋽ was used exclusively as a *mater*, while ⊟ was used exclusively as a letter for writing the /h_/ syllables.

As can be seen from the local scripts of archaic Greece with five vowel letters, the letter O was used to represent the vowels /o(:)/ or /ɔː/. As has just been said, there was no Phoenician sign whose name could rhyme with /no(:)/ or /nɔː/.The Greeks would not be able to find a suitable *mater* to rhyme with /no(:)/ or /nɔː/ in the same way as they would find the *mater* ⋽ to rhyme with /ne(:)/ or /nɛː/. The Phoenicians could not have used the sign O *'ayin* to write any Greek syllables because no Greek syllables would begin with a sound close to the Phoenician pharyngeal fricative /ʕ/. How could the Greeks use this redundant sign as a *mater* to rhyme with /no(:)/ or /nɔː/?

The Phoenicians would probably pronounce the name of the sign O as /ʕajin/ or the like, but the Greeks, having no such sound as /ʕ/ in their phonological system, would probably pronounce it as either /haiin/ or /aiin/. Neither the full name (/(h)aiin/) nor its initial syllable (/(h)ai/) seems to be able to rhyme easily with syllables ending in /o(:)/ or /ɔː/. How could the Phoenician sign O be used as the new rhyming *mater*?

C. Brixhe, a French expert on ancient Greek, suggests that the Greeks probably found the sign O for the vowel /o/ in this way. The name of the Phoenician sign O /ʕajin/ means 'eye' in Phoenician. Probably because the sign looked like an eye, the Greeks called it the eye sign. As it so happened that all the words for 'eye' in ancient Greek began with /o/, the name of the eye sign would begin with /o/ whichever Greek word was chosen. By applying the acrophonic principle, the Greeks could then use the sign O to represent the initial sound of its name: /o/ (2007:285). Brixhe's idea seems to be supported by such ancient Greek words as οφθαλμος /opʰtʰalmos/ 'eye', ομμα /omma/ 'eye', οφις /opsis/ 'sight', οσσε /osse/ 'two eyes', which all begin with the sound /o/.

Brixhe's suggestion about the provenance of the vowel letter Ο is plausible. However, while Brixhe's concern is with how the Phoenician sign Ο could become a vowel letter, our concern here is how it could become a *mater*. In searching for a *mater* for rhyming with syllables ending in /o(:)/ or /ɔː/, some Greek or Greeks might have hit upon the idea of exploiting the initial CV or V syllable of the Greek name of a sign rather than its original Phoenician name. In this case the Greeks should be able to isolate the initial V syllable /o/ of the Greek name of the eye sign and use it to rhyme with syllables ending in /o(:)/. The eye sign Ο, read as /o/ as a *mater*, could then require the preceding sign to rhyme with it. This idea would catch on in the Greek world as it met the need for a much required *mater* that could rhyme with syllables ending in /o(:)/.

It seems rather easy for the Greeks to have found two more *matres*. But it should be noted that the original three *matres* evolved naturally from ancient Egyptian writing, whereas the Greeks' search for two more *matres* represents a deliberate effort to create them. The Greeks must have fully grasped the rhyming principle of the *matres* before they could create two more *matres* by exploiting either the name of a sign or the initial CV or V syllable of its name.

12.4 Early Greek alphabetic writing with five *matres*

Nobody knows exactly when the Greeks created two more *matres*. But from the inscription on the Dipylon vase we know that these two *matres* had already been created by 740 BC. As was said earlier, the Greeks might have been motivated to learn writing with three *matres* from the Phoenicians as early as the tenth century BC. It would probably take the Greeks a long time to develop two more *matres* for common use in their writing to reach the stage at which they had five *matres* at their disposal: Α, Ꙇ, Υ, Ⴁ, and Ο.

At the earliest stage of proto-Greek alphabetic writing, the percentage of literate Greeks should be very low, and the areas to which writing could be applied very limited. The development of proto-Greek alphabetic writing would be slow, only to pick up pace at a later date when the literacy rate went up and the areas of the application of writing widened. Therefore it might have taken a century for proto-Greek alphabetic writing with three *matres* to evolve into early Greek alphabetic writing with five *matres*. The Greeks might have found two more *matres* in the ninth century BC.

We call the stage of Greek writing with five *matres* the beginning of early Greek alphabetic writing, as distinct from proto-Greek alphabetic writing with three *matres*. With five *matres*, the Greeks would be able to write and read Greek syllables more accurately than before. A sign plus a *mater* would basically represent one definite Greek syllable comprising a consonant plus a monophthong, as the diphthongs occurred much less frequently than the monophthongs in ancient Greek (see Appendix 9). This can be seen in Table 19 below, which shows the most common and the less common Greek sound values of ⋏ plus a *mater* in those Greek regions where five *matres* were in use.

Table 19 The Greek sound values of ⋏ plus a *mater* in early Greek alphabetic writing

⋏ plus a *mater*	Most common sound values	Less common sound values
A⋏	/na(ː)/	/na(ː)i/, /nau/
?⋏	/ni(ː)/	-
Y⋏	/nu(ː)/	-
∃⋏	/ne(ː)/	/nɛː/, /nei/, /neu/, /nɛːi/
O⋏	/no(ː)/	/nɔː/, /nou/, /noi/, /nɔːi/, /nui/

It should be noted that in regions where the monophthongs /ɛː/ and /ɔː/ existed, ∃⋏ also represented /nɛː/ apart from /ne(ː)/, and O⋏ also represented /nɔː/ apart from /no(ː)/. Even so, ∃⋏ and O⋏ probably represented /ne(ː)/ and /no(ː)/ much more often than /nɛː/ and /nɔː/ (see Appendix 9).Thus the statement that a sign plus a *mater* in early Greek alphabetic writing basically represented one definite Greek syllable is generally true for all Greek regions where five *matres* were used. Since a sign plus a *mater* in early Greek alphabetic writing basically represented one definite Greek syllable, and since three or even four of the five rhyming *matres* were read as V syllables, early Greek alphabetic writing was in a good position to evolve into a segmental writing system.

How would the Greeks read the five *matres* A, ?, Y, ∃, and O? As can be seen from Table 20 below, in Phoenician the signs ⟨, ⟩, and Y would be read respectively as /ʔa/, /ji/, and /wu/ when used as *matres*. When borrowed by the Greeks for use as *matres*, they would be read, however, as /a/, /i/, and possibly /wu/. The *mater* ⟨

would be read as /a/, because the glottal stop /ʔ/ did not exist in Greek. When the Greeks heard the Phoenician sound /ʔa/, they would tend to miss the initial glottal stop /ʔ/ under the influence of their phonological system. The *mater* ᛝ would be read as /i/, not /ji/, because it is generally believed that the consonant /j/ had been lost early in ancient Greek. It is assumed that the letter ᛝ that appears before another vowel letter in the local scripts of archaic Greece represents the vowel /i(ː)/, not the consonant /j/. While the Phoenicians probably read the name of the sign ᛝ as /joːd/, the Greeks probably read it as /i ɔː ta/ in conformity with their phonological system. As regards the relationship between the consonant /j/ and the vowel /i/, it should be noted that /j/, when lengthened slightly, easily becomes a vocalic /i/ and that /i/, when shortened slightly, easily becomes a consonantal /j/. As for the *mater* Υ, it would probably be read as /wu/ in regions where /w/ could be used as a consonant before a vowel. The sound /wu/ might have evolved into /u/ or /hu/ in regions where /w/ had disappeared before a vowel. It should be noted, however, that /u/, used as a word-initial syllable, is not attested in those archaic Greek inscriptions that are published in Jeffery's *The Local Scripts of Archaic Greece*. We assume therefore that Υ might be read as /hu/ when used as a *mater* in regions where /w/ had disappeared. How the Greeks would read their new-found Greek *matres* ∃ and O has been dealt with in §12.3. The Greeks would probably read the *mater* ∃ as /(h)eː/ or /(h)ɛː/, and the *mater* O as /o/.

Table 20 A comparison of *matres* in Phoenician and early Greek alphabetic writings

	Matres in Phoenician writing				**Matres in early Greek alphabetic writing**			
	Sign	Name	Normal use: syllabic sign representing	Special use: *mater* read as		Sign	Name	Normal use: syllabic sign representing	Special use: *mater* read as
1.	Ҡ	ʾāleph	/ʔ_/	/ʔa/	1.	A	alpha	V syllables	/a/
2.	ᛝ	yōd	/j_/	/ji/	2.	ʓ	iōta	-	/i/
3.	Υ	wāw	/w_/	/wu/	3.	Υ	wau	/w_/	/wu/
					4.	∃	(h)ē	-	/(h)eː/ or /(h)ɛː/
					5.	O	eye sign	-	/o/

As has just been explained, of the five Greek *matres*, A, ʓ, and O were most probably read as V syllables (/a/, /i/, and /o/) in Greek. ∃ was also read as a V syllable (/eː/ or /ɛː/) in many Greek regions. When three or four out of five *matres* were read as V syllables, they would tend to transform themselves from *matres* into vowel letters in early Greek alphabetic writing (this point will be dealt with later on).

When A, ?, ?, and O were on their way to becoming vowel letters, the remaining *mater* Y would tend to change into a vowel letter in an analogous manner, and its reading would thus change from /wu/ into /u/.

The five Phoenician signs ?, ?, Y, ?, and O evolved into A, ?, Y, ?, and O in early Greek alphabetic writing. It should be noted that of the five Greek signs, ?, Y, ?, and O would be used exclusively as *matres*. In Phoenician writing, no matter whether ?, ?, and Y were used as ordinary letters or as *matres*, their shapes remained unchanged. The Greeks, however, wanted to differentiate the shape of an ordinary letter from that of a *mater* in order to facilitate ease of reading. From their experience of using *matres*, the Greek would come to realize that when a sign was used exclusively as a *mater*, it could be identified easily. Once it was identified, its function as a *mater* would become clear and distinct, and this would make reading easier.

It is quite obvious that ? and O would be used only as *matres*, but not as ordinary letters. As there were no such syllables as /j_/ and /ʕ_/ in ancient Greek, ? and O were not required to denote these sounds and so could be used exclusively as *matres*. As regards the letter Y, it was probably used as both an ordinary sign and a *mater* initially. Later, while its variant form ? was used as an ordinary sign standing for /w_/, Y was probably used exclusively as a *mater*, which was read possibly as /wu/. The use of ? exclusively as a *mater* has been discussed at some length earlier in this book. It is possible that in early Greek alphabetic writing ? was used exclusively as a *mater* while ? was used as an ordinary letter for writing the /h_/ syllables.

The case of A, however, is more complicated. As explained in §11.3, the Phoenicians might use ??, ??, and Y? to write the Greek syllables /a/, /i/, and /u/. When the Phoenicians explained to the Greeks that the first sign ? should be read as /ʔa/, /ʔi/, /ʔu/, etc, and that the second sign was a *mater* that determined its sound values, the Greeks would tend to miss the Phoenician glottal stop /ʔ/ under the influence of their mother tongue and would hear /ʔa/, /ʔi/, /ʔu/ as /a/, /i/, /u/. The Greeks would come to understand that the sign ? could be used as either an ordinary letter or a *mater*. When used as an ordinary letter, it should be read as /a/, /i/, /u/, etc; when used as a *mater*, it would indicate that the preceding sign should be read as a syllable ending in /a(:)/.

Proto-Greek alphabetic writing would evolve into early Greek alphabetic writing when two more *matres* were created. When five *matres* were ready for use to write everyday words, Greek writing, presumably, had come of age.

12.5 How the Greek signs changed into vowel and consonant letters

Over time, the five *matres* A, ≀, Y, ∃, and O would evolve into vowel letters. The earliest extant Greek inscription on the Dipylon vase is already a mature piece of segmental writing. The five signs A, ≀, Y, ∃, and O used in the inscription, we believe, are vowel letters, no longer *matres*.

We believe that a *mater* plays an indispensable role in proto-Greek and early Greek alphabetic writings. While the Phoenicians used *matres* sparingly in Phoenician writing, the Greeks would use *matres* to write Greek as a rule. A sign plus a *mater* would be a basic unit of writing in Greek. The reading of a sign in Phoenician writing was context-dependent, whereas that of a sign plus a *mater* in Greek writing would become more definite when the number of *matres* had increased from three to five. When there were only three *matres*, a sign plus a *mater* could stand for five to nine syllables (see Table 17). When five *matres* were in use, a sign plus a *mater* basically stood for one single syllable (see Table 19). For example, ≀И and YИ would be read definitely as /ni(:)/ and /nu(:)/ respectively; AИ, ∃И, and OИ would be read quite often as /na(:)/, /ne(:)/, and /no(:)/ respectively. This is a major departure from Phoenician writing, in which an ordinary sign, written generally without a *mater*, stood for several syllables.

That a sign plus a *mater* basically stood for only one syllable would enable the Greeks to see clearly the relationship between the written form and its pronunciation, which would ultimately lead to the perception of a *mater* as a vowel letter. The five *matres* must have gone through a long period of gestation before they changed into vowel letters. The evolution of the *matres* into vowel letters can roughly be divided in four stages as described in the following paragraphs and as schematically presented in Table 21 below. When reading the following description of the evolution of the *matres*, one may refer to the relevant stages in Table 21.

The written form ≀И is used as an example to illustrate how a *mater* would gradually be perceived as a vowel letter by the Greeks. We will attempt to see the change of the *mater*'s role from the Greeks' perspective when they still had no phonemic concept. Since Greek alphabetic writing was read mainly from right to left before the sixth century BC, the written form ≀И (written as ᒣᒉ in proto-Greek alphabetic writing) is to be read from right to left too. When written *boustrophedon*, ≀И became MS. MS should then be read from left to right.

At stage 1, when there were only three *matres*, 𐤍, still in its Phoenician form, could have as many as twenty-two "sounds". Each sound was the smallest speech sound unit, which one calls a syllable today. The *mater* 𐤉 would reduce the twenty-two sounds of 𐤍 to eight, which can be transcribed today as /ni(:)/, /ne(:)/, /nɛ:/, /nei/, /nɛ:i/, /neu / (see Table 17 above and Table 21.1 below). Any one sound of 𐤍 would be regarded by the Greeks as an integral whole sound unit and is represented here by a light grey rectangle in Table 21.1. The sound of the Greek *mater* 𐤉, transcribed today as /i/, is represented by a dark grey rectangle. Since 𐤍𐤉 had eight sounds, the *mater* 𐤉 would serve as a rough phonetic indicator only. In a piece of writing, the context would further determine which sound of 𐤍𐤉 should be used.

As can be seen from Table 21.1 below, even though 𐤍𐤉 could stand for /ni(:)/, the Greeks were unlikely to regard 𐤉 as representing the last part of the /ni(:)/ sound, because 𐤍𐤉 could also stand for /ne(:)/, /nɛ:/, /nei/, /nɛ:i/, /neu / apart from /ni(:)/. Besides, it stood more often for /ne(:)/ than for /ni(:)/ (see Appendix 9).

Table 21.1 The Greeks' perception of the functions of 𐤍 and 𐤉 at stage 1

Sign	Rectangle
𐤍	/na na: ni ni: nu nu: ne ne: no no: nɛ: nɔ: nei nɛ:i nai na:i nui noi nɔ:i nau neu nou/
𐤉	/i/
𐤍𐤉	/ni ni: ne ne: nɛ: nei nɛ:i neu/

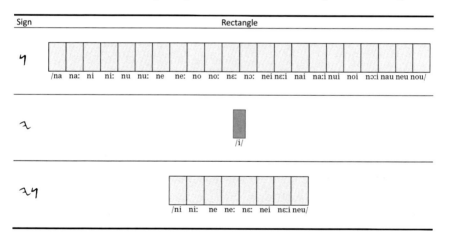

At the outset of stage 1, the pace of development of proto-Greek alphabetic writing must have been very slow as the Phoenicians only needed to write Greek names sporadically. It would also take the Greeks a considerable period of time to learn the Phoenician method of writing Greek sounds and to put it to practical use. The insufficiency of this method would be more and more keenly felt by the Greeks when Greek alphabetic writing was put to wider use. It would also take the Greeks a long time to find two more *matres* to write Greek more accurately. Stage 1 should span a very long period of time.

At stage 2 (see Table 21.2 below), there would be five *matres*. 𐤉 would still have as many sounds as before. However, the *mater* 𐤆 would reduce the number of 𐤉's sounds from twenty-two to only two, transcribed today as /ni/ and /niː/ (see Table 19 above). Thus 𐤆𐤉 would basically have only one sound value: /ni/. That 𐤆𐤉 had only one sound value would be critical in the development of Greek alphabetic writing: a sign plus a *mater* would basically represent a definite sound. When coming across the signs 𐤆𐤉, the Greeks would not have to think about what other sounds 𐤆𐤉 might represent apart from /ni/. That is, they could rule out at once the possibilities that 𐤆𐤉 might also be read as /ne(ː)/, /nɛː/, etc. It would eventually dawn on them that 𐤆𐤉 was to be read unambiguously as /ni/ and only as /ni/. That 𐤆𐤉 had only one sound value is a major departure from Phoenician writing, in which a sign had a set of sound values.

Table 21.2 The Greeks' perception of the functions of 𐤉 and 𐤆 at stage 2

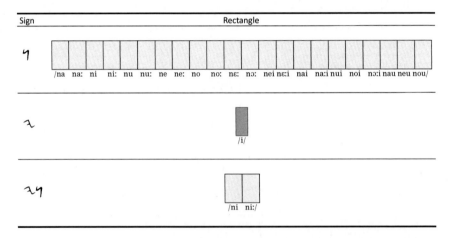

Sign	Rectangle
𐤉	/na naː ni niː nu nuː ne neː no noː nɛː nɔː nei nɛːi nai naːi nui noi nɔːi nau neu nou/
𐤆	/i/
𐤆𐤉	/ni niː/

Stage 3 is the most crucial stage in the evolution of Greek alphabetic writing. It witnesses the transitional period during which Greek alphabetic writing evolved from a syllabic writing system into a segmental one, and it will be dealt with at greater length here. We assume that at this stage the above 𐤆𐤉 would evolve in shape into 𐤆ᴎ. At the beginning of this stage, when coming across the written form 𐤆ᴎ, the Greeks would read it automatically as /ni/. To the Greeks, /ni/ was an integral sound unit, which they would not bother to analyse. However, from their aural and visual experience of reading and writing 𐤆ᴎ and other written forms ending in 𐤆, the Greeks would sooner or later come to realize that not only could the sound of the *mater* ᒿ rhyme with the sound of ᒿᐱ but it was also virtually the same as the rear part of ᒿᐱ's sound. With respect to sound quality, sonority, and duration, the rear part of ᒿᐱ's sound can be said to be nearly the same as the sound of the *mater* ᒿ. Thus it would not be very difficult for the Greeks to identify the rear part of ᒿᐱ's sound with the sound of the *mater*. This being the case, the Greeks might then wonder whether the *mater* ᒿ could be regarded as representing the rear part of ᒿᐱ's sound (see Table 21.3 below).

Table 21.3 The Greeks' perception of the functions of ᐱ and ᒿ at stage 3

Sign	Rectangle
ᒿᐱ	/ni/
ᒿ	/i/
ᒿᐱ	/in/

(the phonemic transcription here is to be read from right to left, to follow the main direction of archaic Greek alphabetic writing)

Traditionally, the Greeks would regard the *mater* ᒿ in ᒿᐱ as a sound indicator of ᐱ only. While ᐱ by itself could have quite a lot of sounds, ᒿ's function was to point out which ones of the sounds that ᐱ should have. ᒿ would not be regarded as a sign that could represent a part of one of ᐱ's sounds. However, when ᒿᐱ basically had only one sound and when the sound of ᒿ happened to be nearly the same as the rear part of this sound, it was then possible for the Greeks to perceive ᒿ as a sign that could represent the rear part of ᒿᐱ's sound.

The visual form <ᒿᐱ> would also make it easier for the Greeks to perceive the sound of ᒿᐱ as comprising two parts—the front and the rear. As the second sign <ᒿ> would gradually come to be perceived by the Greeks as representing the rear part of the sound, the Greeks might begin to wonder whether the first sign <ᐱ> could be regarded as representing the front part of the sound.

The Greeks' perception of the function of the *matres* would begin to change as a result of their experience of reading and writing those written forms ending in ⟨ such as ⟨Ν /ni(:)/, ⟨ᴃ /bi(:)/, ⟨↥ /gi(:)/, ⟨⊿ /di(:)/, etc. Each of these written forms comprises two signs. The second sign ⟨, common to each of these forms, would come to be perceived as representing the rear part of the sound of each written form, as has just been explained. This change in the Greeks' perception of the function of the *mater* ⟨ would make the Greeks wonder whether the function of the other *matres* A, Y, Ⅎ, and O could be perceived in the same way.

As can be seen from Table 19, AΛ and OΛ would be read quite often as /na(:)/ and /no(:)/. When they were read in this way, the Greeks, aided by their perception of the *mater* ⟨ as representing the rear part of a sound, might also come to regard the *matres* A and O (read as /a/ and /o/) as representing the rear parts of /na(:)/ and /no(:)/. This way of perceiving the function of the *matres* A and O in AΛ and OΛ would be reinforced by the Greeks' experience of reading and writing those written forms ending in A and in O, such as Aᴃ, A↥, A⊿, Oᴃ, O↥, O⊿, etc.

It can also be seen from Table 19 that ⅎΛ would be read quite often as /ne(:)/ and YΛ definitely as /nu(:)/. Just as the *matres* ⟨, A, and O could be regarded as representing the rear parts of /ni(:)/, /na(:)/ and /no(:)/ respectively, so the *matres* Ⅎ and Y, by analogy, could be regarded as representing the rear parts of /ne(:)/ and /nu(:)/ respectively, even though the *matres* Ⅎ and Y were not read as V syllables in many Greek regions.

From their experience of reading and writing AΛ, ⟨Λ, YΛ, ⅎΛ, and OΛ, the Greeks would be more and more sure that the *matres* A, ⟨, Y, Ⅎ, and O could be regarded as representing the rear parts of the sounds /na(:)/, /ni(:)/, /nu(:)/, /ne(:)/, and /no(:)/. This way of perceiving the function of the *matres* would be reinforced by the Greeks' experience of reading and writing other written forms ending in a *mater*. Consequently, the five *matres* A, ⟨, Y, Ⅎ, and O would eventually be regarded by the Greeks as representing a part of a syllable. This seems to be a small step taken by the Greeks in the development of their alphabetic writing. However, by taking this small step, the Greeks had in fact taken a giant leap for mankind. Not until the Greeks had taken this step did man realize that a sign could be used to represent a segment of a syllable.

The Greeks could regard a *mater* as representing a part of a syllable because Greek alphabetic writing had evolved into a stage at which the following two conditions were met: (1) two successive signs were used to represent basically a single CV syllable; (2) the second sign was read as a V syllable, which happened to rhyme with the sound represented by the two successive signs. These two conditions do not seem to be very unusual to people used to segmental writing, but it had taken more than two thousand years for these two conditions to arise naturally as a result of the script development from ancient Egyptian writing, through proto-Semitic, proto-Canaanite and Phoenician alphabetic writings, to early Greek alphabetic writing with five *matres*. Had any one of the above writing systems not been created, the above two conditions would not have occurred. It should be noted that before the advent of early Greek alphabetic writing, a *mater* had always been perceived as an indicator of a whole sound, but not as a sign that could represent a component part of a whole sound.

In early Greek alphabetic writing with five *matres*, the above-mentioned two conditions had arisen and so the Greeks would gradually come to realize that a *mater* could be regarded as representing the rear part of a sound. If a *mater* was thus regarded, then the question that inevitably followed would be: could the ordinary sign preceding the *mater* be regarded as representing the front part of the sound? For example, in the case of ⟩M, if the *mater* ⟩ was regarded as representing the rear part of the sound of ⟩M, could the ordinary sign M be regarded as representing its front part? It must be noted that the Greeks had probably never imagined that M could be regarded as representing only a part of a sound because they had been used to perceiving M as representing quite a lot of sounds. Before they thought it possible to regard ⟩ as representing the rear part of the sound of ⟩M, they might even never have thought that the sound of ⟩M could have a front part.

Without the assistance of the visual form ⟨⟩M⟩, it might not have occurred to the Greeks that the sound of ⟩M could have two parts—the front and the rear. If the sounds of ⟩M and ⟩ had been written in Japanese *kana* as に and い, which have the sound values of /ni/ and /i/ respectively, they would probably have regarded the sounds of に and い simply as two different sounds, though with the same rhyme. The written forms of the two Japanese syllabograms would not have given the Greeks any visual clues that enabled them to come to the idea that a syllable could be made up of two parts. If they had regarded the sounds of ⟩M and ⟩ simply as two different sounds, they would not have bothered to give further thought to the relationship between them. It was probably the written form of ⟩M that forced the Greeks to recognize that the sound of ⟩M could have a front part. As the sign ⟩ was perceived as representing the rear part of the sound of ⟩M, the Greeks would be forced to consider whether M could be perceived as representing its front part.

The Greeks would not be sure at first what sort of sound the sign M could represent if the *mater* ⟩ was regarded as representing the rear part of the sound of ⟩M.

To find out what sort of sound it was, they would be obliged to analyse the sound of 𐌉𐌍, probably by comparing it with the sound of 𐌉. They might reason that if 𐌉 was regarded as representing the rear part of the sound of 𐌉𐌍, then 𐌍 could be regarded as representing the remaining part of the sound, that is, the front part. As this front part of the sound could hardly exist on its own in their language, the Greeks would probably have some difficulty in isolating it from the whole sound. In order to see what sort of sound the front part really was, the Greeks might do the following.

They might read out 𐌉𐌍 as [ni] and the *mater* 𐌉 as [i] to see how the sounds were different. They would find that the two sounds were articulated differently at the beginning, though ending in the same rhyme. They might then try out other sounds whose onset was the same as that of [ni], but whose rhymes were different, such as [na], [nu], [ne], [no]. To this end, they might read out 𐌀𐌍, 𐌉𐌍, 𐌖𐌍, 𐌄𐌍, 𐌏𐌍 as [na], [ni], [nu], [ne], [no]. In saying out these sounds, they would feel the presence of a common front part in each sound. In order to find out the phonetic nature of the front part, which was very short in duration, they might try to prolong this part deliberately by pronouncing, say, [na] as [nnna] to make it aurally more distinct. The Greeks would probably learn from this experiment that it was possible to regard 𐌍 as representing the front part of the sound.

They might also try out other sounds with the same rhyme as [ni], but different onsets, for example, by reading out 𐌉𐌔, 𐌉𐌌, 𐌉𐌋, 𐌉𐌁, 𐌉𐌇 as [si], [mi], [li], [bi], [ki]. When saying out these sounds, they would unmistakably feel the presence of a different front part in each sound. In order to further explore the nature of the front parts of these sounds, they might try to prolong the front parts by pronouncing these sounds as [sssi], [mmmi], [llli], [...bi], [...ki] (the sign ... here denotes the holding up of a sound before its release). They would feel from the sounds they said out that their front parts were articulated in different ways and that some of these front parts could be prolonged more easily than others. They could also hear from the sounds they said out that all these front parts were aurally different and that each front part had its own phonetic distinctiveness. The Greeks would probably learn from this experiment that the sign preceding the *mater* could decidedly be regarded as representing the front part of the sound.

As was said in §11.2, the Phoenicians might have used Sign$_1$ + *Mater*$_1$ + Sign$_2$ to write a Greek CVC syllable. If this was the case, when recording such a Greek name as /nesto:r/, the Phoenicians would probably write it as 𐤍𐤉𐤎𐤕𐤅𐤓, which would evolve, as was explained at great length in this book, into 𐌓𐌏𐌕𐌔𐌄𐌍 in early Greek alphabetic writing. When coming across such a written name, the Greeks would read it as /nesto:r/. As a syllable-final consonant /s/ is easier to prolong than a syllable-initial consonant /s/, the Greeks would be able to pronounce [nesss] more easily than, say, [ssse]. When the consonant /s/ was pronounced [sss] as in [nesss], it could be isolated more easily from the whole sound /nes/. Consequently, the Greeks would be able to better grasp the phonetic nature of the syllable-final consonant /s/ that was represented by the sign ⟨ in the written name ⟨OT⟨ƎⱯ. The Greeks could then identify the less isolable syllable-initial consonant /s/ with the more isolable syllable-final consonant /s/, since the syllable-initial /s/ was phonetically similar to the syllable-final consonant /s/ in many ways. Just as the sign ⟨ in the name ⟨OT⟨ƎⱯ was used to represent the more isolable syllable-final consonant /s/, so the same sign ⟨, as was used in the written form, say, ⟩⟨, could be regarded as representing the less isolable syllable-initial consonant /s/, that is, the front part of the syllable /si/. What is said about the sign ⟨ in ⟨OT⟨ƎⱯ applies to the sign ⟨ in ⟨OT⟨ƎⱯ. The point to make in this paragraph is that the Greeks' perception of the signs ⟨ and ⟨ in, say, ⟩⟨ and ⟩⟨, as representing the front part of a syllable would be reinforced by their perception of the signs ⟨ and ⟨ in ⟨OT⟨ƎⱯ as representing the more isolable syllable-final consonants /s/ and /r/.

To recapitulate, the way in which a CV syllable was written in early Greek alphabetic writing would also make it easier for the Greeks to perceive that a sound was made up of two parts—the front and the rear. For example, that /na/, /ni/, /nu/, /ne/, /no/ were written as AⱯ, ⟩Ɐ, YⱯ, ƎⱯ, OⱯ—the same sign (Ɐ) for the front parts but different signs (A, ⟩, Y, Ǝ, and O) for the rear parts—would make it easier for the Greeks to perceive that the front parts of the sounds were the same while the rear parts were all different. On the other hand, that /si/, /mi/, /li/, /bi/ were written as ⟩⟨, ⟩M, ⟩Λ, ⟩ꓭ—the same sign (⟩) for the rear parts but different signs (⟨, M, Λ, ꓭ) for the front parts—would make it easier for the Greeks to perceive that the front parts of the sounds were all different while the rear parts were the same. In other words, the visual representation of a CV syllable in early Greek alphabetic writing probably also played an important role in the Greeks' perception of its component parts.

The Greeks would gradually come to have the idea that a sound could be regarded as being composed of two parts. For example, in the case of ⟩Ɐ, the ordinary sign Ɐ and the *mater* ⟩ would eventually be regarded as representing the front and the rear parts of the sound /ni/ respectively. Thus the sound of ⟩Ɐ in the last row of Table 21.3 is represented by a rectangle comprising two parts that are different shades of grey—the light grey front part and the dark grey rear part. The rear part is dark grey because the Greeks identified the rear part of the sound of ⟩Ɐ with the sound of the *mater* ⟩, which is represented by a dark grey rectangle in the row just above.

Traditionally, the Greeks had been used to regarding ⟨ as an indicator of the sounds of Ϻ. Now that they perceived the functions of ⟨ and Ϻ in a new way, the nature of Greek alphabetic writing was beginning to change, even though the written form ⟨Ϻ for the sound /ni/ remained unchanged. It would probably take the Greeks some considerable time to come to the conclusion that an ordinary sign plus a *mater* could be regarded as representing the front and the rear parts of a sound.

The longer the Greeks used their alphabetic writing with five *matres*, the more they would be convinced that the *matres* A, ⟨, Y, ⱻ, and O in, say, AϺ, ⟨Ϻ, YϺ, ⱻϺ, and OϺ could be regarded as representing the rear parts of the sounds /na(:)/, /ni(:)/, /nu(:)/, /ne(:)/, and /no(:)/ and that the ordinary sign Ϻ could be seen as representing the front part of each of these sounds. The more experience the Greeks had in reading and writing Ϻ, the better they would be able to grasp the transformed nature of Ϻ and its new relationship with the sound it stood for. What had originally been thought to be an integral sound, say, /ni/ as represented by ⟨Ϻ, was eventually split up by the Greeks into two successive discrete parts or segments—/n/ and /i/, which later came to be called a *consonant* and a *vowel*. The Greeks had been used to perceiving Ϻ in ⟨Ϻ as representing a lot of sounds, each one of which could be easily pronounced; now they used it to represent a part of a sound which could hardly exist on its own in their language. The nature of the sign Ϻ had undergone a drastic change.

It must be noted that it would not be easy for people to perceive a consonant as a component part of a spoken sound without the visual aid of its written representation in segmental writing. This point is easy to understand, but tends to be overlooked by many people. Having learnt their alphabet since early childhood, these people take segmental writing so much for granted that they tend to think that such a syllable as /na/ can naturally be analysed into /n/ and /a/. It might not have occurred to them that this kind of analysis is by no means as natural as they think. As a matter of fact, until Greek segmental writing came into being in the first quarter of the first millennium BC, man was not able to isolate even the V element from a simple CV syllable, not to say the C element. There had been neither vowel letters nor consonant letters before the birth of Greek segmental writing. People who regard the Semitic alphabetic scripts as examples of consonantal writing might argue that in a consonantal writing system like Phoenician, from which Greek segmental writing evolved, a sign denotes a consonant. We have argued at great length in this book that in a so-called consonantal writing system like Phoenician, a sign stands for a set of CV syllables with a common onset followed by variable rhymes, not for a consonant.

We believe that to discover the use of a sign to stand for a consonant is no ordinary achievement. That the ancient Greeks were able to discover the use of such a sign was due to the occurrence of the following two conditions in early Greek alphabetic writing: (1) two successive signs represent a single CV syllable; (2) the second sign, read as a V syllable, rhymes with the CV syllable. As was said earlier in this section, the occurrence of these two conditions was contingent upon a combination of many factors. Had any one of these factors been missing, the above two conditions in the Greek alphabetic script would not have arisen.

At stage 4 (see Table 21.4 below), the Greeks would eventually realize that in the written form of ʔᴧ, ᴧ and ʔ could be regarded as representing the front and the rear parts of ʔᴧ's sound. When ᴧ and ʔ were regarded in this way, they were on their way to becoming a consonant letter and a vowel letter respectively. The ordinary sign ᴧ, instead of denoting a lot of Greek syllables with a common onset, was regarded as representing the front part of ʔᴧ's sound, which is called a consonant today. The *mater* ʔ, instead of serving to denote the sound(s) of ᴧ, was regarded as representing the rear part of ʔᴧ's sound, which is called a vowel today.

In Table 21.4 below the sound of ʔᴧ is represented by a rectangle comprising two parts that are the same shade of light grey. The two parts are the same shade because the functions of the two letters ᴧ and ʔ had changed. Instead of ʔ serving as a *mater* to denote the sounds of ᴧ, both ᴧ and ʔ now directly represented the two segments of the sound of ʔᴧ. In other words, as far as sound representation is concerned, ʔ can be considered to be on a par with ᴧ. Hence the sound segments they represented are the same shade of light grey.

Table 21.4 The Greeks' perception of the functions of ∧ and ⟨ at Stage 4

Sign	Rectangle
⟨∧	/in/ (to be read from right to left)
⟨∧	/i/ /n/
⟨	/i/
∧	/n/

The sound of ⟨ lasts longer than that of ∧, and thus ⟨ is represented by the greater part of the rectangle for the sound /ni/, as shown in the second row of the table. Since ⟨ represents the rear part of the sound, ∧ naturally represents the front part. The sound of ∧, by itself, is less prominent and not as long-lasting as the sound of ⟨. Hence it takes up the lesser part of the rectangle. It should be noted that the Greeks did not have to heed the sound durations of ∧ and ⟨. They did not have to know when exactly the sound of ∧ ended and when the sound of ⟨ began. All they needed to know was that the sound of ⟨ followed the sound of ∧.

Previously, ∧ had been a syllabic sign with multiple sound values, whereas ⟨ had been a *mater*. Now to the Greeks, ⟨ represented the rear part of the sound of ⟨∧. The syllabic sign ∧ would thus be reduced to a sign representing only the front part of the sound. Subsequently ∧ would change from a syllabic sign into a consonant letter.

Our explanation of the transformation of the sign ∧ can be said to be fairly close to what I.J. Gelb means by the principle of reduction (1952: 183). Unfortunately, Gelb did not further elaborate on this point. Possibly partly for this reason, other linguists do not seem to have taken the principle of reduction very seriously.

With the vowel and consonant letters in place, Greek writing had at last transformed from a syllabic system to a segmental one. However, the forms of all the letters, including the *matres*, remained the same throughout the transformation period. Outwardly, nothing seemed to have changed. What had changed was the Greeks' perception of the functions of the letters, which led to a change in the nature of the letters. The change of the inner structure of Greek alphabetic writing from a syllabic system to a segmental one is not easily discerned. This could possibly be one of the reasons why the origin of the Greek alphabet has not been satisfactorily explained for so long. This change of the nature of Greek alphabetic writing is of enormous significance in the history of writing. It has given rise to a new type of writing system—segmental writing. Thanks to Greek segmental writing, mankind was able for the first time to split a syllable into its component parts—a consonant and a vowel.

With five vowel letters and less than twenty consonant letters, the Greeks could write quite accurately all kinds of syllable structure in their language, such as V (including monophthongs and diphthongs), VC, and CCV. They would re-examine the existing written words and, if necessary, re-spell them for a more refined and accurate representation of the sound segments of the words. When writing a word not yet written before, they could analyse its sound *segment by segment* and then write out the segments with the appropriate Greek letters. To be able to write the sound of a word in this way must have been a novel and thrilling experience for the Greeks. Take for example the written representation of the syllable /nai/. In proto-Greek alphabetic writing the Greeks would probably write /nai/ as A⋀, as they had not yet been able to write diphthongs accurately with the *matres* they then had (see Table 14). Now that they could write /nai/ out segment by segment as ⟨A⋀, they must have felt that the new written form ⟨A⋀ was a significant improvement over the old one A⋀ for the syllable /nai/.

13   The creation of new vowel letters in early Greek alphabetic writing

When the five *matres* became vowel letters, what were their sound values in the scripts? If one were to disregard vowel length, Α, Ɩ, and Υ would each represent one vowel quality in all Greek regions while Ⴈ and Ο would each represent two vowel qualities in many Greek regions. In regions where the dialect had the following seven vowel qualities /a, i, u, e, o, ɛ, ɔ/, Ⴈ and Ο would each represent two vowel qualities: Ⴈ would represent both /e/ and /ɛ/, and Ο both /o/ and /ɔ/.

In regions where Ⴈ and Ο each represented two vowel qualities, there was a need to create two more vowel letters so that each vowel letter could neatly represent one vowel quality. However, the need was not so urgent as to make the creation of two more vowel letters obligatory, for a Greek alphabetic script could still function quite well even when Ⴈ and Ο each represented two vowel qualities. /e/ and /ɛ/ could simply be written as Ⴈ, and /o/ and /ɔ/ simply as Ο. Reading Ⴈ and Ο, however, would pose a slight problem. Ⴈ would be read as either /e/ or /ɛ/, and Ο as either /o/ or /ɔ/. As was said in §12.2, the ancient Greeks had a habit of reading aloud. When reading aloud an unfamiliar word spelt with such a letter as Ⴈ or Ο, they might hesitate about the sound value of the letter because the letter had two different sound values. Such instances would make it less convenient for them to read aloud.

Regions speaking a dialect with seven vowel qualities were divided as to whether it was necessary to create two more vowel letters. Some regions created two more vowel letters so that each vowel letter could neatly represent one vowel quality, such as the eastern Ionic Dodekapolis; others were content to stick to five vowel letters for writing their dialect with seven vowel qualities, such as Attica.

According to A. Malikouti-Drachman, Old Attic of the fifth century BC had the following twelve monophthongs: /a(ː), i(ː), u(ː), e(ː), o(ː), ɛː, ɔː/ (2007:526). If vowel length is discounted, Old Attic can be said to have seven vowel qualities. It is also stated in the article on the Greek language in the *Encyclopaedia of Britannica* (15th edition) that the above vowel system "may be attributed to Old Attic of about 500 BC" (2003: vol. 22, p. 615). If Old Attic had seven vowel qualities, then the sound values of the five vowel letters in the Attic inscriptions of archaic Greece were probably as follows: Α /a(ː)/, Ɩ /i(ː)/, Υ /u(ː)/, Ⴈ /e(ː), ɛː/, Ο /o(ː), ɔː/. Ⴈ and Ο each represented two vowel qualities. As a matter of fact, the Athenians kept on using their alphabet with only five vowel letters until 403 BC when they voted to replace it with the eastern Ionic alphabet of Miletus with seven vowel letters. When Attic was written in the Milesian alphabet, the sound values of the seven vowel letters were: Ａ /a(:)/, Ζ /i(:)/, Υ /u(:)/, Ⴈ /e(:)/, Ｏ /o(:)/, Ⱨ /ɛ:/, Ω /ɔ:/. Each vowel letter neatly represented one vowel quality.

The fact that seven vowel letters were used in the inscriptions of the eastern Ionic Dodekapolis suggests that the eastern Ionic dialect most likely had seven vowel qualities. The sixth and seventh vowel letters in the alphabets of the Ionic Dodekapolis were Ⱨ and Ω. How were these two vowel letters created?

13.1 The creation of the sixth vowel letter in Greek alphabetic writing

The sixth vowel letter was derived from the Phoenician sign Ⱨ *ḥēth*. Its name was pronounced /hɛ:ta/ in many Greek regions, but in regions where /h/ had disappeared, it was pronounced /ɛ:ta/. In a region where Ⱨ was called /ɛ:ta/, Ⱨ could be used as a vowel letter to represent /ɛ:/ for the following reason. When early Greek alphabetic writing became segmental, a letter in the Greek alphabet would generally represent the initial sound or segment of its name. For example, the letter Ａ would represent /a(:)/, which is the initial sound or segment of its name *alpha*. This is the so-called acrophonic principle. By applying this principle, the Greeks could then use Ⱨ to represent the vowel /ɛ:/, which is the initial sound or segment of its name *ēta*. Now that Ⱨ represented /ɛ:/, Ⴈ could represent only /e(:)/. While Ⱨ was called /ɛ:ta/, Ⴈ was probably called the /e(:)/ sign.

From Inscriptions 1*a*, 1*a-b*, 1, and 1 on Plates 59, 61, 63, and 67 respectively of *The Local Scripts of Archaic Greece* (Jeffery 1961: at end), one can see that Ⱨ was used as a vowel letter for /ɛ:/ in Thera and in Rhodes as early as the late eighth century BC and in Crete and in Samos more than half a century later in the second half of the seventh century BC. In Thera and in Rhodes the letter Ⱨ, besides representing /ɛ:/, was also used as a consonant letter for /h/, whereas in Crete and in Samos Ⱨ was used exclusively as a vowel letter since /h/ had been lost in the Cretan and Samian dialects. Under these circumstances, it seems more likely that the sixth vowel letter Ⱨ was first created in Crete or Samos than in Thera or Rhodes, because, while one can use the acrophonic principle to account for the creation of Ⱨ as a vowel letter in Crete or Samos, it is more difficult to explain how Ⱨ could evolve into a vowel letter in Thera or Rhodes when it had to represent /h/ at the same time. We believe that it is possible that the sixth vowel letter ⊟ was first created in Crete or in Samos in the eighth century BC even though it is first attested in the Cretan and Samian inscriptions dated to the seventh century BC. In the history of writing it is not unusual that a lapse of time might occur between the creation of a sign in a place and the first attestation of its existence in that place. It is possible that Thera and Rhodes adopted the use of ⊟ as a vowel letter from Crete or Samos in the eighth century BC.

Besides Thera and Rhodes, many other Greek regions used ⊟ to represent both /ε:/ and /h/, such as the central and northern Aegean islands (Ionic), Knidos, and Kleonai. In other words, ⊟ was used both as a vowel letter and as a consonant letter in these regions. Only a few Greek regions went one step further by using ⊟ and its variant form to represent /h/ and /ε:/ respectively. From Inscriptions 31-34 on Plate 68 of *The Local Scripts of Archaic Greece*, one can see that at Knidos, an ancient Greek city not far away from Rhodes, two variant forms of ⊟, namely H and ▯, were used to represent /h/ and /ε:/ respectively in the sixth century BC. And from Inscriptions 5-7 on Plates 24-25 of *The Local Scripts of Archaic Greece*, one can see that at Kleonai, an ancient Greek city not far away from Corinth, the letters ⊟ and ẞ were used to represent /h/ and /ε:/ respectively in the sixth century BC. Can one say that just as H and ▯ were used to represent /h/ and /ε:/ respectively at Knidos, so ⊟ and ẞ were used to represent /h/ and /ε:/ respectively at Kleonai? Is ẞ a variant form of ⊟?

Kleonai probably adopted the use of ẞ as a vowel letter from its powerful neighbour Corinth. In the eighth century BC, Corinth was a flourishing commercial centre. There was a real need for a Greek script to keep administrative and trading records there. The newly-created Greek segmental alphabet that spread from region to region during the eighth century BC must have been most welcome at Corinth when it had arrived there. Judging from the archaic Corinthian inscriptions, the Greek alphabet that Corinth received seems to be an eastern alphabet (see §14). What makes the Corinthian alphabet distinct from the parent alphabet is its use of both ẞ and Ǝ as vowel letters for mid front vowels. ẞ seems to be the sixth vowel letter created by the Corinthians. How was it created? What exactly were the sound values of ẞ and Ǝ?

According to Jeffery, at Corinth ẞ was "employed for normal ε and η" while Ǝ was "used for the false diphthong ει only" (1961:24). By this statement Jeffery probably means that at Corinth ẞ represented /e/ and /ε:/ while Ǝ represented the monophthong /e:/ that had evolved from the diphthong /ei/. The Greek scholar J. M. Dosuna, however, says that at Corinth "standard epsilon represents short [e] and primary long open [ε:]" and that "a special letter (Corinthian epsilon) is used for secondary long close [e:] (resulting from lengthening and contraction, and from the monophthongization of [ei] <ει>)" (2007:447). If by "standard epsilon" and by "a special letter (Corinthian epsilon)" Dosuna means ⊐ and 𝌆 respectively, then Dosuna seems to be saying in his statement that ⊐ represents /e/ and /ɛː/ while 𝌆 represents /eː/. Dosuna's statement about the sound values of ⊐ and 𝌆 seems to be at variance with Jeffery's. Whose statement is correct?

As regards Dosuna's statement, it should be noted that in the Corinthian inscriptions 𝌆 was used much more frequently than ⊐. This means that the vowel(s) represented by 𝌆 occurred much more frequently than the vowel(s) represented by ⊐. This is an important point to bear in mind when one talks about the sound values of 𝌆 and ⊐. It is unlikely that in the Corinthian inscriptions ⊐ represented /e/ and /ɛː/ while 𝌆 represented /eː/, because had this been the case, ⊐ should have occurred much more frequently than 𝌆 in the Corinthian inscriptions.

Regarding Jeffery's statement about the sound values of ⊐ and 𝌆, we have some queries. If it is true that 𝌆 represented /e/ and /ɛː/ while ⊐ represented /eː/, then this implies the following. First, discounting vowel length, 𝌆 represented two vowel qualities (/e/ and /ɛ/) while 𝌆 and ⊐ represented the same vowel quality (/e/). Second, the creation of the sixth vowel letter 𝌆 had not helped to improve Greek spelling because on the one hand the Corinthians would still have to rely on the context to decide whether 𝌆 should be read as /e/ or /ɛː/ and on the other they would have to decide whether to use 𝌆 or ⊐ when writing a sound with the vowel quality /e/. One cannot help wondering why the Corinthians had not made better use of the sixth vowel letter 𝌆 so that 𝌆 and ⊐ could each neatly represent one vowel quality. Is it possible that the short /e/ and the primary long open /ɛː/ differed so little in quality that they were considered by the Corinthians to be the same vowel quality? Should this be the case, the short /e/ could be transcribed as the more open /ɛ/, which is distinct from the secondary long close /eː/ that resulted, as Dosuna says, "from lengthening and contraction, and from the monophthongization of [ei]". If one could reconstruct the vowel system of the Corinthian dialect, one would be able to see more clearly the correlation between its vowels and the six vowel letters in the Corinthian alphabet.

How was the vowel letter 𝌆 created at Corinth? Scholars are divided on this point. Some believe that 𝌆 is a variant of ⊟. Some hold that 𝌆 is a variant of ⊐. Others say that the Corinthians deliberately and unconventionally used a redundant variant of the *bēta* sign as a vowel letter. We now comment on these opinions one by one.

Some scholars attribute the creation of the sixth vowel letter 𝌆 to the employment of a variant of ⊟. The Corinthians probably knew that ⊟, besides representing /h/, was also used as a vowel letter for the mid front vowel in some Greek regions. To disambiguate the function of ⊟, the Corinthians used ⊟ only as a consonant letter for /h/ and its variant 𐤇 only as a vowel letter for the mid front vowel. The first question that arises from this hypothesis is: could 𐤇 be regarded as a derivation of ⊟ in shape? On the face of it they do not look very much alike, but even so, one cannot deny that there is a sort of family resemblance between them and that 𐤇 could have derived from ⊟. The variant of ⊟ could be written as 𐤇 without being mistaken for the *bēta* sign because the *bēta* sign was written in the Corinthian alphabet as Ⴑ, which seems to have derived from the *courtyard* sign 𓉐 in Egyptian hieroglyphics.

Some scholars believe that 𐤇 is a cursive variant of the vowel letter Ⴈ. If this hypothesis is correct, then the sixth vowel letter 𐤇 was an original Corinthian creation. Ⴈ and its variant 𐤇 were used as vowel letters for the mid front vowels at Corinth. The first question that arises from this hypothesis is similar to the one posed in the above paragraph: could 𐤇 be regarded as a derivation of Ⴈ in shape after all?

Some scholars think that the Corinthians unconventionally used a redundant consonant letter as a vowel letter to represent the mid front vowel. The *bēta* sign in the Corinthian alphabet had two forms: Ⴑ and 𐤇. As the Corinthians used only Ⴑ for /b/, 𐤇 became redundant. Since there was a need to create one more vowel letter for the mid front vowel, some Corinthian or Corinthians had the unconventional idea of using 𐤇 as the sixth vowel letter. The other Corinthians soon followed suit. If this hypothesis is correct, then the Corinthians' decision to arbitrarily use a consonant letter as a vowel letter was a bold move in the history of segmental writing.

The alphabets of Corinth's neighbours Megara, Sikyon, and Kleonai are basically identical with that of Corinth. We now focus again on the sixth vowel letter in these alphabets. At Megara as at Corinth, the sixth vowel letter 𐤇 was used much more frequently than Ⴈ, which means that the sound values of 𐤇 and Ⴈ were possibly the same in the two places. By contrast, the sixth vowel letter 𐤇 was used much less frequently than Ⴈ at Kleonai, which means the sound values of 𐤇 and Ⴈ at Kleonai must have been different from those at Corinth and Megara. As regards the sixth vowel letter at Sykion, it was written as 𐌗, not 𐤇. This is probably because the *bēta* sign was written as 𐤇 at Sykion, as in many other places. The Corinthian 𐤇 had to assume another shape at Sykion, which turned it into 𐌗.

13.2 The creation of the seventh vowel letter in Greek alphabetic writing

It is quite certain how and why the seventh vowel letter was created in Greek alphabetic writing. The seventh vowel letter is Ω (or C), which everyone agrees is a variant of O. As was said earlier, in Greek regions speaking a dialect with seven vowel qualities but writing in an alphabet with only five vowel letters, the letter O represented both /o(:)/ and /ɔ:/. To neatly represent these vowels, some of these Greek regions created the seventh vowel letter Ω, such as the eastern Ionic Dodekapolis and three of the Aegean islands Delos, Syros, and Keos. In these regions O represented /o(:)/, and Ω represented /ɔ:/.

It is not so certain, however, when and where the seventh vowel letter was first created in Greek alphabetic writing. By referring the reader to Guarducci's *Epigrafia greca* (1967:101 and 159-60), C. Brixhe says that Ω "appears for the first time, at the turn of the eighth and seventh centuries, on a Parian graffito on Delos" (2007:281). If the dating of the Parian graffito is correct, then one can say that the seventh vowel letter Ω was first created probably in the eighth century BC, even though one cannot say for certain that it was first created on Paros or Delos, or elsewhere.

In Jeffery's *The Local Scripts of Archaic Greece* are collected some of the earliest archaic Greek inscriptions in which the seventh vowel letter Ω is attested, such as Inscriptions 61, 2, and 26 on Plates 58, 63, and 56 respectively. The first one is a marble block inscription of the late seventh century BC from Thasos; the second one is a gravestone inscription of the early sixth century BC from Samos, and the third one a marble column inscription of the first half of the sixth century BC from Paros. It seems that many Greek regions came to use the seventh vowel letter Ω for the first time much later than the eighth century BC.

It should be noted that the seventh vowel letter Ω had different sound values in different Greek regions. As has just been said, in the eastern Ionic Dodekapolis and on three of the Aegean islands Delos, Syros, and Keos, O represented /o(:)/ while Ω represented /ɔ:/. In these regions, O was probably called the /o(:)/ sign, and Ω the /ɔ:/ sign. However, on Thasos and Paros, O and its variant Ω swapped their sound values: O represented /ɔ:/ while Ω represented /o(:)/. In these regions, O was probably called the /ɔ:/ sign, and Ω the /o(:)/ sign. It is possible that the names of O and Ω determined their sound values.

In Melos, Sikinos, Anaphe, and Knidos, the letter O and its variant C (a half circle) were used to represent /ɔ:/ and /o(:)/ respectively. In these regions, O was probably called the /ɔ:/ sign, and C the /o(:)/ sign. In Paros and Thasos, the *bēta* sign happened to be the same shape as C; therefore, to avoid confusion with *bēta*, C, a variant of O, was turned 90° clockwise and, to make it visually more distinct from the other letters, was given two splay feet for it to change into Ω. This could possibly account for the shape of Ω. Just as O and C were used to represent /ɔ:/ and /o(:)/ respectively in Melos, Sikinos, Anaphe, and Knidos, so O and Ω were used to represent /ɔ:/ and /o(:)/ respectively in Paros and Thasos.

14 ## The creation of new consonant letters in early Greek alphabetic writing

As explained in §12.5, when the five *matres* changed into vowel letters, the ordinary Phoenician signs used for writing Greek would inevitably change from syllabic signs into consonant letters. This shift in the nature of Greek alphabetic writing probably took place in the ninth century BC. What were the sound values of the consonant letters of the Greek alphabet when Greek writing had become segmental?

We assume that before Greek alphabetic writing became segmental, the ancient Greeks had already decided to use (1) only ⊗ for /tʰ_/ rather than both ⊗ and T; (2) only T for /t_/ rather than both ⊗ and T and (3) only ⟩ or M (variants of W) for /s_/ rather than W, ‡, and ↜. If this assumption is correct, then when Greek alphabetic writing became segmental, the consonant letters would have sound values as shown in Table 22 below:

Table 22 The sound values of consonant letters when Greek alphabetic writing became segmental

৪	↖	⊿	⅂	I	日	⊗	Ӿ	(	⅍	⋏	⅂	φ	⟨	⟩ or M	T
b	g	d	w	dz	h	tʰ	k, kʰ	l	m	n	p, pʰ	k, kʰ	r	s	t

As can be seen from Table 22 above, although it is generally true that each letter represented one consonant and that each consonant was represented by one letter, there are some exceptions. For instance, the letter ⅂ represented both /p/ and /pʰ/ sounds, while the /k/ and /kʰ/ sounds were each represented by the letters Ӿ and φ. Such instances might pose a slight problem for the ancient Greeks in reading and in writing.

When writing /k/ or /kʰ/, the ancient Greeks had to choose between Ӿ and φ. This, however, would not pose much of a problem for them because there was a simple rule for them to follow regarding the use of Ӿ and φ. When /k/ or /kʰ/ was followed by /u(ː)/ or /o(ː)/, they would use φ; otherwise, they would use Ӿ. This could be the main reason why the ancient Greeks seemed to be quite happy to keep on using both Ӿ and φ for /k/ until the sixth century BC when φ was gradually replaced by Ⴙ (see §11.1.1).

When reading aloud an unfamiliar word spelt with such a letter as ⨅, Ⴙ, or Ϙ, the ancient Greeks might hesitate about the sound value of the letter because the letter had two different sound values: ⨅ represented both /p/ and /pʰ/; Ⴙ represented both /k/ and /kʰ/, and so did Ϙ. Such instances would make it less convenient for the ancient Greeks to read aloud. There was thus a need to represent /p/, /pʰ/, /k/, and /kʰ/ more precisely. Judging from the local scripts of archaic Greece, the ancient Greeks except the Cretans did take the trouble to create more graphemes so as to represent /k/, /kʰ/, /p/, and /pʰ/ more precisely. The Cretans did not seem to mind very much that ⨅, Ⴙ, and Ϙ each had two sound values. They went on using ⨅ for /p/ and /pʰ/ and Ⴙ for /k/ and /kʰ/ until the fourth century BC when all the Greek regions came to adopt the Milesian alphabet as the standard. As for Ϙ, the Cretans used it also for /k/ and /kʰ/, but replaced it with Ⴙ in the sixth century BC.

14.1 The creation of the new digraphs for /pʰ/ and /kʰ/ in Thera

To the ancient Greeks, /p/, /pʰ/, /k/, and /kʰ/ were four different sounds. They might not know, however, how these sounds were different. Judging from the way in which /pʰ/ and /kʰ/ were written in ancient Thera, these sounds had been analysed and found to be aspirated. Inscriptions 1a(i) and 5 on Plate 61 in *The Local Scripts of Archaic Greece* (Jeffery 1961: at end) show that in ancient Thera, while /p/ and /k/ were written as ⨅ and Ⴙ respectively, /pʰ/ and /kʰ/ were written as 日⨅ and 日Ⴙ respectively (for simplicity's sake, Ϙ, the alternative of Ⴙ, is ignored from now on). In 日⨅ and 日Ⴙ, 日 represented the aspirate sound /h/. As Inscription 1a(i) on Plate 61 is dated to the end of the eighth century BC, the Therans had probably learned to write /pʰ/ and /kʰ/ as 日⨅ and 日Ⴙ at a date earlier than 700 BC.

The Theran method of writing /p/, /pʰ/, /k/, and /kʰ/ probably implies the following: (1) As /p/, /pʰ/, /k/, and /kʰ/ were four different sounds, the Therans wanted to represent them differently in writing. (2) When Greek alphabetic writing became segmental, the ancient Greeks would come to realize that different letters generally represented different segments or sounds. By focusing on the sound represented by a consonant letter, they would be able to grasp better and better the nature of this sound. With the accumulated knowledge of the nature of the sounds represented by the Greek consonant letters, the Therans could set about analysing the sounds of /p/ and /pʰ/ to see how they were different. Eventually they would come to realize that /p/ was unaspirated, while /pʰ/ was aspirated. They might reason that if they could write /p/ as ⨅, then they could write /pʰ/ as 日⨅, because the aspirate sound could be represented by the letter 日, which they used for /h/. The Therans hence used ⨅ for /p/

and ⊟˥ for /pʰ/. By analogy, they used ㅐ for /k/ and ⊟ㅐ for /kʰ/. It may be noted that the Therans' method of notating /pʰ/ and /kʰ/ over 2500 years ago is basically the same as the one used by the International Phonetic Association today.

The Theran method of writing /p/, /pʰ/, /k/, and /kʰ/, scientific as it is, has some shortcomings. ⊟˥ could be read as /pʰ/, but it could also be read as /pɛ:/, because ⊟ could also be used in Thera as a vowel letter for /ɛ:/. Thus the Therans had to rely on the context to decide on the sound value of ⊟˥. Similarly, ⊟ㅐ could be read as /kʰ/ or /kɛ:/. It can be said that the Therans' solution to a problem gave rise to another problem. This phenomenon is by no means rare in the evolution of writing.

The use of ⊟˥ and ⊟ㅐ for /pʰ/ and /kʰ/ is also attested in Inscription 23 on Plate 62 dated to the late sixth century BC from Melos, an island not far away from Thera. It seems that the Theran method of writing /pʰ/ and /kʰ/ had spread to Melos during the seventh or sixth century BC.

14.2 The creation of two more letters for /pʰ/ and /kʰ/

While Thera managed to represent /p/, /pʰ/, /k/, and /kʰ/ differently in writing without creating a new letter, the eastern and western Greeks created two more letters so that these four sounds could each be neatly represented by a distinct letter. To neatly represent /p/, /pʰ/, /k/, and /kʰ/, the eastern Greeks used ˥, Φ, ㅐ, and Χ while the western Greeks used ˥, Φ, ㅐ, and ⌄ , respectively.

To represent /pʰ/, both the eastern and western Greeks exploited Φ, which we believe is a variant of the Phoenician sign φ qōph. That so many different regions arbitrarily employed the same sign Φ for /pʰ/ probably bespeaks a single source. Because of the arbitrary relationship between the letter Φ and the sound /pʰ/, the numerous Greek city-states could not have created Φ to represent /pʰ/ independently of each other.

When Greek writing became segmental, the Greeks would come to regard a consonant letter as representing the first consonant or sound of its name. For example, the letter ၓ represented /b/, which is the first sound of its name bēta. The name bēta originates from the name of the Phoenician sign ৭ bēth. The name bēth, meaning 'house' in Phoenician, reminds the Phoenicians that the sign ৭ originally depicts a house. Even though ৭ is not very pictorial, its shape is somehow bound by the meaning of its name bēth as it evolves. However, when the name became bēta in Greek, it had lost its original meaning. To the Greeks, bēta was just the name of a letter. The shape of the letter *bēta* was thus free to evolve in Greek writing, so long as it remained distinct from the other letters. Thus generally speaking, the relation between the shape of a Greek letter and the sound it represented became more and more arbitrary.

It is possible, therefore, that when some Greek had the bold or unconventional idea of using Φ for /pʰ/, the other Greeks did not find it objectionable in principle to use Φ in this way. What bothered them would be that the new letter Φ should look distinct from Ϙ *qōppa*. Once they were assured that the two letters were distinct, the idea of using Φ for /pʰ/ would catch on as it had met the need for a new letter for /pʰ/. The Greeks would then give the new letter a name with the initial sound /pʰ/, say, *phi* /pʰi/. The Greeks would come to realize that any new letter, when given a name that complied with the acrophonic principle, could be used to represent the initial sound of that name as long as it was distinct from the existing letters. In other words, the choice of this letter could be arbitrary. This was no ordinary discovery in the history of writing, which one tends to lose sight of today. Equipped with this piece of knowledge, the Greeks could, in theory, write any Greek phoneme that did not exist in Phoenician, be it a vowel or a consonant.

As can be seen from the archaic Greek inscriptions published in Jeffery's *The Local Scripts of Archaic Greece*, the use of Φ for /pʰ/ was widespread in ancient Greece. To find out where and when the new letter Φ was first created, one can only resort to the earliest inscriptions of the Greek regions in which Φ is used for /pʰ/ and hazard a guess. To this end, some relevant bits of information about seven such inscriptions are listed in Table 23 below:

Table 23 Seven of the earliest Greek inscriptions in which Φ is used for /pʰ/

Inscription No.	Plate No.	Date of Inscription	Place of Inscription
2	1	8th century ?	Attica
1	45	c. 700 ?	Ithake
1	47	c. 700 ?	The Euboic colonies, Italy
1	7	c. 700-675 ?	Boiotia
4	18	c. 675-650 ?	Corinth
2	55	c. 650 ?	Naxos
1	63	c. 650-600 ?	Samos

Judging from the above table, the letter Φ was first created for /pʰ/ probably in the eighth century BC, possibly in Attica in Central Greece. If this hypothesis is correct, one may say that the idea of using Φ for /pʰ/ spread from Attica both eastwards and westwards and won the approval of the Greeks everywhere except those living on the southernmost islands in the Aegean Sea such as the Cretans and Therans.

As for the representation of /kʰ/, the eastern and western Greeks employed different letters. The eastern Greeks arbitrarily chose X or ✝, which we believe is a variant of T *tau*, while the western Greeks used Ѵ, which we believe is a variant of Ⴗ *kaph*. The new letter X or Ѵ used for /kʰ/ would be given an appropriate name, say, *khi* /kʰi/, and with an appropriate name X or Ѵ could be used to represent /kʰ/.

The use of X for /kʰ/ is analogous with the use of Φ for /pʰ/. Just as the letter Φ used for /pʰ/ had to be distinct from the letter Ϙ used for /k(h)/, so the letter X used for /kʰ/ had to be distinct from the letter T used for /t/.

The use of Ѵ for /kʰ/ is analogous in a sense with the use of Ω for /ɔː/. Just as the vowel letter O and its variant Ω were used to represent /o(ː)/ and /ɔː/ respectively, so the consonant letter Ĥ and its variant Ѵ were used to represent /k/ and /kʰ/ respectively.

To find out where and when X was first used for /kʰ/, one has to resort to the earliest Greek inscriptions in which X is used for /kʰ/. We refer again to some of the earliest inscriptions published in *The Local Scripts of Archaic Greece*. Some relevant bits of information about four such inscriptions are listed in Table 24 below:

Table 24 Four of the earliest Greek inscriptions in which X is used for /kʰ/

Inscription No.	Plate No.	Date of Inscription	Place of Inscription
1	1	c. 725 ?	Attica
1 a-b	18	c. 700 ?	Corinth
2	55	c. 650 ?	Naxos
2	63	c. 600-575 ?	Samos

Judging from Table 24, X was first used for /kʰ/ probably in the eighth century BC, possibly in Attica.

While the choice of X for /kʰ/ is quite arbitrary, the choice of Ѵ for /kʰ/ is not completely arbitrary. Ѵ, a variant of Ѵ, was possibly used to represent the Greek /k_/ and /kʰ_/ at a very early date. This letter had probably become redundant when Greek writing became segmental. As there was a need to represent /k/ and /kʰ/ more precisely in Greek writing, the idea of employing this redundant letter for /kʰ/ might come to some Greek. When he used Ѵ to write /kʰ/, the other Greeks would probably find it a good idea to write /kʰ/ in this way and so would follow suit.

Where and when was Ѵ first used for /kʰ/? Again we resort to some of the earliest inscriptions published in *The Local Scripts of Archaic Greece*. See Table 25 below:

Table 25 Six of the earliest Greek inscriptions in which Ѵ is used for /kʰ/

Inscription No.	Plate No.	Date of Inscription	Place of Inscription
1	7	c. 700-675 ?	Boiotia
3 a	67	7th century ?	Rhodes
2	44	c. 625 ?	Aitolia
1	12	c. 600-550 ?	Phokis
1	11	c. 550 ?	Thessaly
9 A 1-2	5	c. 550-525 ?	Euboia

Judging from Table 25, ⍦ was first used for /kʰ/ probably in the eighth century BC, possibly in Boiotia. The idea of using ⍦ for /kʰ/ possibly spread from Boiotia to the neighbouring regions such as Euboia, Thessaly, Aitolia, and Achaia.

14.3 The creation of two more letters for /ks/ and /ps/

How the ancient Greeks represented the consonant clusters /ks/ and /ps/ in writing is also noteworthy. When Greek writing became segmental, each of these clusters could well be represented by two consonant letters—one letter for /k/ or /p/, and the other for /s/. It is small wonder that so many Greek regions used two consonant letters to represent them. Some regions used Ħ for /k/, ⌐ for /p/, and M for /s/, while some others used ⋏ for /kʰ/, Φ for /pʰ/, and ⟩ for /s/. It seems that the two consonant clusters were perceived as /ks/ and /ps/ by some Greeks and as /kʰs/ and /pʰs/ by others. Consequently, the clusteres were written as M Ħ and M ⌐ in some regions and as ⋊⋏ and ⟩Φ in others. It is possible that the clusters were spoken either as [ks] and [ps] or as [kʰs] and [pʰs]. [kʰs] and [pʰs] were possibly a more formal way of speaking than [ks] and [ps]. When the clusters were spoken as [ks] and [ps], the aspirate sound was elided to reduce the effort of speaking.

While many Greek regions used two letters to represent /ks/ and /ps/, many other regions, however, went one step further by arbitrarily choosing one single new letter for each cluster. For example, the Ionic Dodekapolis arbitrarily chose Ξ and ⍦ for /ks/ and /ps/ respectively, while Thessaly arbitrarily chose ✝ for /ks/. Each of the new letters would be given an appropriate name. Ξ and ✝ might be called *ksi* /ksi/, and ⍦ *psi* /psi/.

The local scripts of archaic Greece can be classified according to how /p/, /pʰ/, /k/, /kʰ/, /ks/, and /ps/ are written. This is basically the idea of the German classical scholar A. Kirchhoff, who labelled the three types of scripts thus classified as "green", "blue" and "red" on a map of Greece. The "green" type can further be divided into "dark green" and "light green", and so can the "blue" type. Table 26 below shows the five types of Greek script, the localities where the types were used, and the ways in which /p/, /pʰ/, /k/, /kʰ/, /ks/, and /ps/ were written:

Table 26 How /p/, /pʰ/, /k/, /kʰ/, /ks/, and /ps/ are written in the five types of Greek scripts

Type	Locality	/p/	/pʰ/	/k/	/kʰ/	/ks/	/ps/
Dark Green	Crete	ᒣ	ᒣ	ꓘ : Ϙ	ꓘ : Ϙ	Mꓘ	Mᒣ
Light Green	Thera, Melos, Sikinos, Anaphe	ᒣ	⊟ᒣ	ꓘ : Ϙ	⊟ꓘ : ⊟Ϙ	Mꓘ	Mᒣ
Dark Blue	Dodekapolis, Corinth, Argos	ᒣ	Φ	ꓘ : Ϙ	X : +	Ξ	Ψ
Light Blue	Attica, Boiotia, Aigina	ᒣ	Φ	ꓘ : Ϙ	X : +	ΣX	ΣΦ
Red	Thessaly, Phokis, Lokrides, Elis	ᒣ	Φ	ꓘ : Ϙ	Ψ	X : +	ΣΦ

As can be seen from the above table, /pʰ/, /kʰ/, /ks/, and /ps/ are written differently in different Greek regions. We now use four concrete examples to illustrate this point. If the sound of the English word *pea* [pʰi:] were to be written in archaic Greek letters in different Greek regions, it would be written like this: ʔᒣ in Crete, ʔ⊟ᒣ in Thera, and ΙΦ in most other city-states (it may be noted that the same letter may assume different shapes in different regions). The sound of the English word *key* [kʰi:] would be written like this: ʔꓘ in Crete, ʔ⊟ꓘ in Thera, ΙX in Attica, Corinth, and Miletus, and ΙΨ in Thessaly. The sound of the English word *leaps* [li:ps] would be written like this: Mᒣʔᑕ in Crete and Thera, Ψʔᑕ in Corinth and Miletus, and ΣΦʔᑕ in Attica, Boiotia, and Thessaly. The sound of the English word *sex* [seks] would be written like this: MꓘƎM in Crete and Thera, ΞBM in Corinth, ΞƎΣ in Miletus and Samos, ΣXƎΣ in Attica, and XƎΣ in Thessaly.

One can see the diversity of the local scripts of archaic Greece from the above four examples. Thus the city-states needed to adopt a standard alphabet if they wanted to communicate with one another more efficiently in writing. From around 400 BC onwards, the Greek city-states gradually came to adopt the Milesian alphabet as the standard alphabet. By 350 BC the Milesian alphabet had evolved into the classical Greek alphabet of 24 letters that was used in common by all the Greek city-states. The names of the letters all conform to the acrophonic principle: a letter in the alphabet basically represents the initial phoneme or sound of its name. The classical Greek alphabet is nearly phonemic, as one letter basically represents one phoneme.

Table 27 below is a comparative table on the sound values of the signs in four alphabets—the Phoenician alphabet of the tenth century BC, a hypothetical Greek alphabet of the tenth century BC, the Attic alphabet of the sixth century BC, and the classical Greek alphabet of 350 BC which evolves from an eastern Ionic alphabet.

By comparing the sound values of the signs in one alphabet in Table 27 with those of the corresponding signs in the alphabet in the next column, one can see how in essence Greek alphabetic writing developed in the first millennium BC, and one can come to the following conclusions.

First, even though the hypothetical Greek alphabet of the tenth century BC is in essence of the same nature as the Phoenician alphabet, there are some important differences. One important difference is that a Phoenician sign represents far fewer CV syllables than the corresponding Greek letter. Another important difference is that *matres* are used much more persistently and regularly in Greek writing than in Phoenician writing.

Second, even though the letters in the hypothetical Greek alphabet of the tenth century BC and the Attic alphabet of the sixth century BC are more or less the same in outward form, they are in fact drastically different in nature. While a letter in the former alphabet may represent as many as twenty-two CV syllables with a common onset, a letter in the Attic alphabet basically represents only a single vowel or a single consonant.

Third, the classical Greek alphabet of 350 BC with seven vowel letters (highlighted in the table) represents a vocalic system with seven vowel qualities more precisely than the Attic alphabet with five vowel letters (also highlighted in the table).

We believe that among all the local alphabets of archaic Greece, the Cretan alphabet is the closest to the parent alphabet of the ninth century BC when Greek alphabetic writing first became segmental.

Table 27 **A comparative table on the sound values of the signs in the Phoenician alphabet and the corresponding letters in three Greek alphabets of different eras**

Phoenician alphabet of tenth century BC		*Greek alphabet of tenth century BC		Attic alphabet of sixth century BC		Classical Greek alphabet of 350 BC	
Sign	Sound	Sign	Sound	Sign	Sound	Sign	Sound
𐤀	/ʔ_/	𐤀	Mater or V syllalbes	A	/a(:)/	A	/a(:)/
𐤁	/b_/	𐤁	/b_/	𐌁	/b/	B	/b/
𐤂	/g_/	𐤂	/g_/	𐌂	/g/	Γ	/g/
𐤃	/d_/	𐤃	/d_/	𐌃	/d/	Δ	/d/
𐤄	/h_/	𐤄	/h_/	𐌄	/e(:), ɛ:/	E	/e/
𐤅	/w_/	𐤅	mater or /w_/	𐌅	/w/	--	--
𐤆	/z_/	𐤆	/dz_/	𐌆	/dz/	Z	/zd/
𐤇	/ħ_/	𐤇	/h_/	𐌇	/h/	H	/e:/
𐤈	/tˤ_/	𐤈	/t⁽ʰ⁾_/	𐌈	/tʰ/	Θ	/tʰ/
𐤉	/j_/	𐤉	mater	?	/i(:)/	Ι	/i(:)/
𐤊	/k_/	𐤊	/k⁽ʰ⁾_/	𐌊	/k/	K	/k/
𐤋	/l_/	𐤋	/l_/	𐌋	/l/	Λ	/l/
𐤌	/m_/	𐤌	/m_/	𐌌	/m/	M	/m/
𐤍	/n_/	𐤍	/n_/	𐌍	/n/	N	/n/
𐤎	/s_/	𐤎	/s_/	--	--	Ξ	/ks/
𐤏	/ʕ_/	--	--	O	/o(:), ɔ:/	O	/o/
𐤐	/p_/	𐤐	/p⁽ʰ⁾_/	𐌐	/p/	Π	/p/
𐤑	/sˤ_/	𐤑	/s_/	--	--	--	--
𐤒	/q_/	𐤒	/k⁽ʰ⁾_/	𐌒	/k/	--	--
𐤓	/r_/	𐤓	/r_/	𐌓	/r/	P	/r/
𐤔	/ʃ_/	𐤔	/s_/	𐌔	/s/	Σ	/s/
𐤕	/t_/	𐤕	/t⁽ʰ⁾_/	T	/t/	T	/t/
				Υ	/u(:)/	Y	/y(:)/
				Φ	/pʰ/	Φ	/pʰ/
				X	/kʰ/	X	/kʰ/
						Ψ	/ps/
						Ω	/ɔ:/

Concluding remarks

15.1 The importance of writing foreign names in the history of writing

The importance of writing foreign names in the history of writing cannot be exaggerated. The writing of the sounds of the names in a foreign language is basically the writing of the sounds of that foreign language. Usually attention is focussed just on the sound of a name, not its meaning. The form of writing that is built on the writing of foreign sounds may be crude at the outset; however, a fully-fledged writing system may, over time, develop from it. One good example is the birth of Greek alphabetic writing.

The method of writing foreign names in ancient Egyptian and in Phoenician writings plays a pivotal role in the birth of Greek alphabetic writing. Since the ancient Egyptians needed to write foreign names by means of the so-called monoconsonantal signs with multiple sound values, *matres* were created in their writing to specify the sound values of these signs. The Phoenicians probably inherited the Egyptians' method of using *matres* to write foreign names. When this method was applied to the writing of everyday Greek words, the groundwork for Greek alphabetic writing had been laid.

15.2 The Phoenicians' role in the writing of Greek names

The choice of certain Phoenician signs for the representation of some Greek sounds suggests that the Phoenicians probably led the way in writing Greek sounds at the initial stage of Greek alphabetic writing. The Greeks must have followed the Phoenicians' lead in using ⅄ for Greek syllables comprising /k$^{(h)}$/ plus a front vowel and φ for Greek syllables comprising /k$^{(h)}$/ plus a back vowel. If the Greeks had taken the initiative in writing their /k_/ syllables, they would have used the same sign (probably ⅄) for all of them. That Ɪ was chosen to represent /dz/ may also suggest the Phoenicians' role in the writing of Greek. As Phoenician had no such sound, it is little wonder that the Phoenicians would miss the initial [d] sound and perceive /dz/ as /z/. As the Phoenicians used the sign Ɪ for /z_/, they would use this sign for the Greek /dz_/. If the Greeks had taken the initiative in writing /dz_/, they might have chosen different signs.

As was said earlier, the Phoenicians needed to record Greek in their trading contacts with the Greeks, most likely Greek names. If this was the case, *matres* would be used persistently to write Greek names. The Phoenicians would never dream that their method of writing Greek names could lay the basis for Greek alphabetic writing.

15.3 The current views on the genesis of the Greek alphabet

The genesis of the Greek alphabet has not been satisfactorily explained for a number of reasons, the main ones of which can be summed up as follows.

First, not a Greek name is attested in the extant Phoenician inscriptions dated to the first quarter of the first millennium BC, nor the use of *matres* for that matter. As a result, nobody knows for certain how Greek names were written in Phoenician in those days. However, it would be too hasty to jump to the conclusion that the Phoenicians did not need to record Greek names in their writing. The above-mentioned Phoenician inscriptions are so few and they are used for such limited purposes that it is not surprising that Greek names written in Phoenician are not attested. We have argued that the Phoenicians were roving traders who would record Greek names with both ordinary signs and *matres* on papyrus for book-keeping. Papyrus being a perishable writing material, these trading records may be lost to us for ever. If one assumes that *matres* were not used in the Phoenician script, one will have great difficulty in explaining the genesis of Greek alphabetic writing.

Second, there is no scriptorial evidence of Greek alphabetic writing dating before the second half of the eighth century BC. Consequently, nobody knows for certain how the ancient Greeks wrote in the Phoenician alphabet in the first quarter of the first millennium BC. However, if one assumes that there was no Greek alphabetic writing during this period of time, one will have great difficulty in explaining why the earliest extant Greek alphabetic inscriptions on the Dypilon vase and the Nestor's Cup dating from the second half of the eighth century BC are such mature pieces of Greek segmental writing.

Third, for lack of scriptorial evidence of Greek alphabetic writing of the tenth and ninth centuries BC, nobody knows for certain how Phoenician signs were used to write Greek at the initial stages. The genesis of the Greek alphabet is generally explained nowadays like this. The Phoenician script is a consonantal writing system. A Phoenician sign can thus be regarded as representing a consonant. When the Greeks learned the Phoenician alphabet from the Phoenicians, they learned to regard a Phoenician sign as representing a consonant. So they used the Phoenician signs to represent the Greek consonants. When some Phoenician signs, which could also serve as *matres*, turned into vowel letters through the acrophonic principle, the Greek alphabet became a "true" alphabet. Some letters in the Greek alphabet represented Greek vowels, and the others Greek consonants.

To find out whether the Phoenicians would see the Phoenician signs as representing the Phoenician consonants and how the *matres* actually function in the ancient scripts,we trace the ancestry of the Phoenician alphabet all the way back to Egyptian hieroglyphic writing in this book, and by so doing we have come to these conclusions. The Phoenicians would not see a Phoenician sign as representing a Phoenician consonant. Instead, they would see it as having several sounds, which, in linguistic terms, are basically several CV syllables with a common onset followed by different rhymes. As a Phoenician sign had several sounds, the Phoenicians, in writing an unfamiliar foreign name, would use an appropriate *mater* after the Phoenician sign to indicate which one of its sounds was intended. The *mater* required the preceding Phoenician sign to rhyme with it. A *mater* is thus a syllable indicator, not a vowel indicator. A *mater* is not a vowel letter either, as it is mute in its role as a syllable indicator. The Phoenicians probably inherited this method of writing foreign names from the ancient Egyptians.

15.4 The paradigm shift from syllabic writing to segmental writing

It was a great achievement for mankind to be able to break up a syllable into segments termed vowels and consonants today. It took a script like Phoenician to encounter a language like Greek for this to happen. When the Phoenicians used three *matres* to write Greek and when the Greeks developed their writing on this basis with the addition of two more *matres*, the consonant and vowel letters would sooner or later come into being. The concepts of *consonant* and *vowel* seem to have arisen naturally from a series of contingent factors, not from any super-intelligent human design. While syllabic writing was created independently many times in the history of writing, segmental writing was created independently once and once only.

As a Greek CV syllable was denoted as a norm by two letters in proto-Greek alphabetic writing—a syllabic sign and a *mater*, when the *mater*, which originally specified the sound value of the syllabic sign, turned into a vowel letter, the syllabic sign was reduced to a consonant letter. These two letters gave birth to the concepts of *vowel* and *consonant*. It is only the visual representation of a syllable with two letters that forced the Greeks to consider the real nature of the two letters. Only then could the Greeks begin to understand what a vowel and a consonant were. Paradoxical as it may sound, the concepts of *vowel* and *consonant* can be said to originate from writing, not from phonetic analysis.

To the best of our knowledge, ours is a new hypothesis about the origin of the Greek alphabet. It can explain how the Greek alphabet naturally evolved from the Phoenician signs, without having to assume the prior existence of consonantal writing before Greek segmental writing was created. That is, it can explain the paradigm shift from a syllabic writing system to a segmental one without having to go through the intermediary stage of a so-called consonantal writing system.

References

Allen, W. Sidney. 1974. *Vox Graeca: A Guide to the Pronunciation of Classical Greek*. Great Britain: Cambridge University Press.

Asher, R.E. & J.M.Y. Simpson (eds.). 1994. *The Encyclopedia of Language and Linguistics*. Great Britain: Pergamon Press.

Aubet, Maria E. 2001. *The Phoenicians and the West*. 2nd edn. Trans. M. Turton. Cambridge: Cambridge University Press.

Brixhe, Claude. 2007. History of the Alphabet: Some Guidelines for Avoiding Oversimplification. In A.-F. Christidis (ed.), 277-287.

Chadwick, John. 1994. Greek, Ancient. In Asher & Simpson (eds.), vol. 3, 1493-1495.

Chao, Yuen-ren. 1968. *Language and Symbolic Systems*. Cambridge: Cambridge University Press.

Christidis, A.-F. (ed.). 2007. *A History of Ancient Greek: From the Beginnings to Late Antiquity*, Cambridge & New York: Cambridge University Press.

Coulmas, Florian. 1989. *The Writing Systems of the World*. Great Britain: Blackwell.

Coulmas, Florian. 2003. *Writing Systems: An Introduction to Their Linguistic Analysis*. UK: Cambridge University Press.

Culican, William. 1986. Phoenicians. In *The Encyclopedia Americana*, vol. 21, 947-952. USA: Grolier Inc.

Daniels, Peter T. 1996. The Study of Writing Systems. In Daniels & Bright (eds.), 3-17.

Daniels, Peter T. & William Bright (eds.). 1996. *The World's Writing Systems*. U.S.A.: Oxford University Press.

Davies, W.V. 1987. *Egyptian Hieroglyphs*. London: British Museum Publications.

DeFrancis, John. 1989. *Visible Speech: The Diverse Oneness of Writing Systems*. Honolulu: University of Hawaii Press.

Diringer, David. 1962. *Writing*. London: Thames and Hudson.

Driver, G.R. 1948. *Semitic Writing: From Pictograph to Alphabet*. London: Oxford University Press.

Gardiner, Alan H. 1957. *Egyptian Grammar: Being an Introduction to the Study of Hieroglyphs*, 3rd edn. Oxford: Oxford Griffith Institute.

Gaur, Albertine. 1984. *A History of Writing*. London: The British Library.

Gelb, Ignace J. 1952. *A Study of Writing*. USA: University of Chicago Press.

Gnanadesikan, Amalia E. 2009. *The Writing Revolution: Cuneiform to the Internet*. Wiley-Blackwell.

Gordon, Cyrus H. 1965. *Ugaritic Textbook: Grammar*. Rome: Pontifical Biblical Institute.

Havelock, Eric A. 1976. *Origins of Western Literacy*. Canada: The Ontario Institute for Studies in Education.

Healey, John F. 1990. *The Early Alphabet*. Great Britain: British Museum.

Healey, John F. 1994. Alphabet: Development. In Asher & Simpson (eds.), vol.1, 74-79.

Jeffery, Lilian H. 1961. *The Local Scripts of Archaic Greece*. Oxford: Clarendon.

Lejeune, Michel & Cornelis Jord Ruijgh. 2003. Greek Language. In *The New Encyclopaedia Britannica*, 15ᵗʰ edn., vol. 22, 612-615. U.S.A.: Chicago

Loprieno, Antonio. 1995. *Ancient Egyptian: A Linguistic Introduction*. Great Britain: Cambridge University Press.

Malikouti-Drachman A. 2007. The Phonology of Classical Greek. In A.-F. Christidis (ed.), 556-570.

Mieroop, Marc Van De. 2011. *A History of Ancient Egypt*. Wiley-Blackwell.

Naveh, Joseph. 1987. *Early History of the Alphabet*. Jerusalem: The Hebrew University, The Magnes Press.

Olson, David R. 2003. Writing. In *The New Encyclopaedia Britannica*, 15ᵗʰ edn., vol. 29, 1025-1034. U.S.A.: Chicago

Petrounias, Evangelos B. 2007. The Pronunciation of Classical Greek. In A.-F. Christidis (ed.), 556-570.

Powell, Barry B. 2009. *Writing: Theory and History of the Technology of Civilization*. Wiley-Blackwell.

Robins, R.H. 1971. *General Linguistics: An Introductory Survey*. Longman.

Sáenz-Badillos, Angel. 1993. *A History of the Hebrew Language*. Trans. John Elwolde. Great Britain: Cambridge University Press.

Sampson, Geoffrey. 1985. *Writing Systems: A Linguistic Introduction*. London: Hutchinson.

Sass, Benjamin. 1988. *The Genesis of the Alphabet and Its Development in the Second Millennium B.C.* Munich: Wiesbaden.

Simpson, J.M.Y. 1994. Writing Systems: Principles and Typology. In Asher &

Simpson (eds.), vol. 9, 5052-5061.

Swiggers, Pierre. 1996. Transmission of the Phoenician Script to the West. In Daniels & Bright (eds.), 261-270.

Teffeteller, A. 2006. Greek, Ancient. In *The Encyclopedia of Language and Linguistics*, ed. Keith Brown, 2nd edn., vol. 5, 149-153. UK: Elsevier.

Voutiras, Emmanuel. 2007. The Introduction of the Alphabet. In Christidis (ed.), 556-570.

Appendix 1

The monoconsonantal signs in the Egyptian hieroglyphic script*

Sign	Transliteration	Sound Values	Object Depicted
	ꜣ	R_ > ?_	vulture
	i	j_ > ?_	reed
	y	j_	two reeds or oblique strokes
	'	ʕ_	forearm
	w	w_	quail chick
	b	b_	foot
	p	p_	stool
	f	f_	horned viper
	m	m_	owl
	n	n_	water
	r	r_	mouth
	h	h_	reed shelter
	ḥ	ħ_	twisted flax
	ḫ	x_	placenta(?)
	ẖ	ç_	animal's belly with teats
	z	z_	bolt
	s	s_	folded cloth
	š	ʃ_	pool
	ḳ	q_	hill slope
	k	k_	basket with handle
	g	g_	stand for jars
	t	t_	loaf
	ṯ	ʧ_	tethering rope
	d	d_	hand
	ḏ	ʤ_	snake

* See §7.2.2 and A. Loprieno (1995:15)

Appendix 2

A chart comparing four Semitic alphabetic scripts

	Proto-Semitic	Ugaritic			Phoenician			Hebrew			Arabic		
	Sound Values	Sign	T.	Sound Values	Sign	T.	Sound Values	Sign	T.	Sound Values	Sign	T.	Sound Values
1	ʔ_	⟨sign⟩	a	ʔa(:)	⟨sign⟩	ʾ	ʔ_	א	ʾ	ʔ_, ɸ_	ا	ʾ	ʔ_
2	b_	⟨sign⟩	b	b_	⟨sign⟩	b	b_	ב	b,v	b_, v_	ب	b	b_
3	ɡ_	⟨sign⟩	g	ɡ_	⟨sign⟩	g	ɡ_	ג	ɡ,ǧ	ɡ_, ʤ_	ج	ǧ	ʤ_
4	x_	⟨sign⟩	ḫ	x_	/x/ > /ḥ/ *			/x/ > /ḥ/ *			خ	ḥ	x_
5	d_	⟨sign⟩	d	d_	⟨sign⟩	d	d_	ד	d	d_	د	d	d_
6	h_	⟨sign⟩	h	h_	⟨sign⟩	h	h_	ה	h	h_	ه	h	h_
7	w_	⟨sign⟩	w	w_	⟨sign⟩	w	w_	ו	v	v_	و	w	w_
8	z_	⟨sign⟩	z	z_	⟨sign⟩	z	z_	ז	z,ž	z_, ʒ_	ز	z	z_
9	ḥ_	⟨sign⟩	ḥ	ḥ_	⟨sign⟩	ḥ	ḥ_	ח	ḥ	ḥ_	ح	ḥ	ħ_
10	tˤ_	⟨sign⟩	ṭ	tˤ_	⟨sign⟩	ṭ	tˤ_	ט	ṭ	t_	ط	ṭ	tˤ_
11	j_	⟨sign⟩	y	j_	⟨sign⟩	y	j_	י	y	j_	ي	y	j_
12	k_	⟨sign⟩	k	k_	⟨sign⟩	k	k_	כ	k,kh	k_, x_	ك	k	k_
13	ʃ_	⟨sign⟩	š	ʃ_	⟨sign⟩	š	ʃ_		š	ʃ_	ش	š	ʃ_
14	l_	⟨sign⟩	l	l_	⟨sign⟩	l	l_	ל	l	l_	ل	l	l_
15	m_	⟨sign⟩	m	m_	⟨sign⟩	m	m_	מ	m	m_	م	m	m_
16	ð_	⟨sign⟩	ḏ	ð_	/ð/ > /z/ *			/ð/ > /z/ *			ذ	ḏ	ð_
17	n_	⟨sign⟩	n	n_	⟨sign⟩	n	n_	נ	n	n_	ن	n	n_
18	zˤ_	⟨sign⟩	ẓ	ðˤ_	/ðˤ/ > /sˤ/ *			/ðˤ/ > /sˤ/ *			ظ	ẓ	ðˤ_
19	s_	⟨sign⟩	s	s_	⟨sign⟩	s	s_	ס	s	s_	س	s	s_
20	ʕ_	⟨sign⟩	ʿ	ʕ_	⟨sign⟩	ʿ	ʕ_	ע	ʿ	ʕ_, ɸ_	ع	ʿ	ʕ_
21	p_	⟨sign⟩	p	p_	⟨sign⟩	p	p_	פ	p,f	p_, f_	ف	f	f_
22	sˤ_	⟨sign⟩	ṣ	sˤ_	⟨sign⟩	ṣ	sˤ_	צ	ṣ,č	ts_, ʧ_	ص	ṣ	sˤ_
23	q_	⟨sign⟩	q	q_	⟨sign⟩	q	q_	ק	q	k_	ق	q	q_
24	r_	⟨sign⟩	r	r_	⟨sign⟩	r	r_	ר	r	r_	ر	r	r_
25	θ_	⟨sign⟩	t	θ_	⟨sign⟩	t > š	θ > ʃ	ש	š,ś	ʃ_, s_	ث	t	θ_
26	ɣ_	⟨sign⟩	ġ	ɣ_	/ɣ/ > /ʁ/ *			/ɣ/ > /ʁ/ *			غ	ġ	ɣ_
27	t_	⟨sign⟩	t	t_	⟨sign⟩	t	t_	ת	t	t_	ت	t	t_
28	dˤ_	/dˤ/ > /sˤ/ *			/dˤ/ > /sˤ/ *			/dˤ/ > /sˤ/ *			ض	d	dˤ_
29	ɬ_	/ɬ/ > /ʃ/ *			/ɬ/ > /ʃ/ *			/ɬ/ > /s/ *				ɬ > ʃ *	
		⟨sign⟩	i	ʔi(:)									
		⟨sign⟩	u	ʔu(:)									
		⟨sign⟩	ś	su									

Points to note:

(1) *T.* in the table stands for *Transliteration*.

(2) The symbol ϕ in the table stands for a sound that has disappeared.

(3) Proto-Semitic is only a hypothetical language thought to be the common ancestor of all Semitic languages. Presumably it had 29 consonantal phonemes, which have been nearly all inherited by modern Arabic, as is shown in the table above.

(4) The Proto-Canaanite alphabet that had spread to Ugarit most probably had at least 27 signs, from which the first 27 signs in the Ugaritic alphabet originate. The Phoenician alphabet, despite being a direct offshoot of the Proto-Canaanite alphabet, had only 22 signs. The decrease in the number of signs in the Phoenician alphabet is probably due to a decrease in the number of consonantal phonemes in the Phoenician language. When a dialect of Proto-Canaanite with 27 consonantal phonemes evolved into Phoenician in the last quarter of the second millennium BC, five consonants /x, θ, ð, ðˤ, ɣ/ had shifted backwards in the oral cavity and merged respectively with /ħ, ʃ, z, sˤ, ʕ/. As a result, Phoenician had only 22 consonantal phonemes. Thus the signs that had originally stood for /x, θ, ð, ðˤ, ɣ/ became redundant. The redundant signs, except the one for /θ_/, became obsolete. The sign for /θ_/, namely W, was used to represent /ʃ_/, and the original sign for /ʃ_/ was abandoned instead. See §9.4 of this book.

(5) The ordering of the signs in the above table is determined by that of the Ugaritic alphabet. The ordering of the Phoenician and Hebrew alphabets is essentially the same as that of the Ugaritic alphabet. The ordering of the Arabic alphabet, however, is different from that of the Ugaritic alphabet, because the Arabic letters have been reshuffled.

Appendix 3
Proto-Semitic consonant chart

Place of Articulation / Manner of Articulation		Bilabial	Labio-dental	Dental	Alveolar	Post-alveolar	Palatal	Velar	Uvular	Pharyngeal	Glottal
Nasal		m			n						
Stop	Plain	p b			t d			k g	q		ʔ
	Emphatic				tˤ dˤ						
Fricative	Plain			θ ð	s z	ʃ		x ɣ		ħ ʕ	h
	Emphatic				sˤ zˤ						
Lateral Fricative					ɬ						
Trill					r						
Approximant							j	w			
Lateral Approximant					l						

Remarks:

(1) The Proto-Semitic language is generally believed to have 29 consonants.

(2) Ugaritic, a Semitic language akin to Phoenician, has 27 consonants. The consonant /zˤ/ in Proto-Semitic evolved into /ðˤ/ in Ugaritic. The consonants /dˤ/ and /ɬ/ merged with /sˤ/ and /ʃ/ respectively in Ugaritic.

(3) The five consonants /x, θ, ð, ðˤ, ɣ/ in Proto-Canaanite would later shift respectively to /ħ, ʃ, z, sˤ, ʕ/ in Phoenician.

Appendix 4
Ugaritic consonant chart

Place of Articulation / Manner of Articulation		Bilabial	Labio-dental	Dental	Alveolar	Post-alveolar	Palatal	Velar	Uvular	Pharyn-geal	Glottal
Nasal		m			n						
Stop	Plain	p b			t d			k g	q		ʔ
	Emphatic				tˤ						
Fricative	Plain			θ ð	s z	ʃ		x ɣ		ħ ʕ	h
	Emphatic			ðˤ	sˤ						
Trill					r						
Approximant							j	w			
Lateral Approximant					l						

Appendix 5
Phoenician consonant chart

Place of Articulation / Manner of Articulation		Bilabial	Labio-dental	Dental	Alveolar	Post-alveolar	Palatal	Velar	Uvular	Pharyn-geal	Glottal
Nasal		m			n						
Stop	Plain	p b			t d			k g	q		ʔ
	Emphatic				tˤ						
Fricative	Plain				s z	ʃ				ħ ʕ	h
	Emphatic				sˤ						
Trill					r						
Approximant							j	w			
Lateral Approximant					l						

Appendix 6
Modern Hebrew consonant chart

Manner of Articulation / Place of Articulation	Bilabial	Labio-dental	Dental	Alveolar	Post-alveolar	Palatal	Velar	Uvular	Pharyngeal	Glottal
Nasal	m			n						
Stop	p b			t d			k g			ʔ
Fricative		f v		s z	ʃ ʒ			ʁ	ħ ʕ	h
Affricate				ts	ʧ ʤ					
Trill										
Approximant						j				
Lateral Approximant				l						

Appendix 7
Modern Arabic consonant chart

Manner of Articulation / Place of Articulation		Bilabial	Labio-dental	Dental	Alveolar	Post-alveolar	Palatal	Velar	Uvular	Pharyngeal	Glottal
Nasal		m			n						
Stop	Plain	b			t d			k	q		ʔ
	Emphatic				tˤ dˤ						
Fricative	Plain		f	θ ð	s z	ʃ		x ɣ		ħ ʕ	h
	Emphatic			ðˤ	sˤ						
Affricate						ʤ					
Trill					r						
Approximant							j	w			
Lateral Approximant					l						

Appendix 8

Archaic Greek consonant chart

Place of Articulation / Manner of Articulation	Bilabial	Labio-dental	Dental	Alveolar	Post-alveolar	Palatal	Velar	Uvu-lar	Pharyn-geal	Glottal
Nasal	m			n						
Stop	p^h p b			t^h t d			k^h k g			
Fricative				s						h
Affricate				dz						
Trill				r						
Approximant						(j)	w			
Lateral Approximant				l						

(1) The consonant /j/ before a vowel seems to have been lost in archaic Greek dialects. The letter ∑ before a vowel letter in archaic Greek inscriptions seems to denote the vowel /i(:)/ rather than the consonant /j/.

(2) /w/ still existed in most Greek dialects, denoted by Ⅎ in archaic Greek inscriptions.

(3) The affricate /dz/ is regarded as a C phoneme here.

Appendix 9
Frequencies of vowel letters in archaic Greek inscriptions

A mini-survey was conducted to find out the frequency of occurrence of the vowel letters (including digraphs for diphthongs) in archaic Greek inscriptions. Inscriptions dated the 7th and 6th centuries BC from six Greek regions were selected from the relevant plates of inscriptions from these regions in Jeffery's *The Local Scripts of Archaic Greece* and used as samples for the survey. As the purpose of this survey was to find out the relative frequencies of the vowel letters used in archaic Greek inscriptions, consonant letters were not counted. Only vowel letters were counted. The counting of letters was based on the transliteration of plates on pp. 401, 404, 405, 413, and 414 of Jeffery's book. It was found that the frequencies of the vowel letters ∃ and O generally amounted to about 20% each. From the findings, it is apparent that the mid vowels were among the most frequently used vowels in ancient Greek.

In the box below, the first row lists out the vowel letters used in archaic Greek inscriptions, and the second row shows their corresponding transliteration in modern Greek letters. It should be noted that archaic Greek inscriptions could run either from right to left or vice versa and that the same letter could appear in different shapes in different Greek regions. The letter shapes in the first row in the box below are typical, and the digraphs in the first row should be read from right to left.

A	≀	Y	∃	O	⊟	Ω	≀A	≀∃	≀O	≀⊟	≀Ω	YA	Y∃	YO	YΩ	≀Y
α	ι	υ	ε	ο	η	ω	αι	ει	οι	ηι	ωι	αυ	ευ	ου	ωυ	υι

In each of the six tables below, the first row shows the transliteration of the vowel letters in archaic Greek inscriptions. The second row shows the number of occurrences of each vowel letter, and the third row its frequency in percentage.

1. Miletus

A total of five Miletus inscriptions dated the 6th centuries BC taken from Plate 64 (Nos. 23, 26-27, 29, 33) in Jeffery's *The Local Scripts of Archaic Greece* were used as samples. The findings of the vowel letter frequencies are listed in Table 1 below.

Table 1 Frequencies of vowel letters in the inscriptions from Miletus

α	ι	υ	ε	ο	η	ω	αι	ει	οι	ηι	ωι	υι	αυ	ευ	ου	ωυ
35	23	7	35	29	14	10	6	3	9	5	3	0	0	7	2	0
18.6	12.2	3.7	18.6	15.4	7.4	5.3	3.2	1.6	4.8	2.6	1.6	0	0	3.7	1.1	0

Total number of occurrences of the vowel letters: 188

Presumably the seven monographs <α, ι, υ, ε, ο, η, ω> represent /ɑ(ː), i(ː), u(ː), e(ː), o(ː), εː, ɔː/ respectively, and the ten digraphs <αι, ει, οι, ηι, ωι, υι, αυ, ευ, ου, ωυ> represent /ɑi, ei, oi, εːi, ɔːi, ui, au, eu, ou, ɔːu/ respectively.

As can be seen from Table 1 above, the vowel letters <α, ε, ο, ι> occurred much more frequently than the others. This could mean that among all the Greek vowels, /ɑ(ː), e(ː), o(ː), i(ː)/ occurred the most frequently.

2. Samos

A total of six Samos inscriptions dated the 7th and 6th centuries BC taken from Plate 63 (Nos. 1-2, 4-5, 8, 13) in Jeffery's *The Local Scripts of Archaic Greece* were used as samples. The findings of the vowel letter frequencies are listed in Table 2 below.

Table 2 Frequencies of vowel letters in the inscriptions from Samos

α	ι	υ	ε	ο	η	ω	αι	ει	οι	ηι	ωι	υι	αυ	ευ	ου	ωυ
18	9	4	13	12	16	4	0	0	0	3	1	0	0	1	0	0
22.2	11.1	4.9	16	14.8	19.8	4.9	0	0	0	3.7	1.2	0	0	1.2	0	0

Total number of occurrences of the vowel letters: 81

As can be seen from Table 2 above, the vowel letters <α, η, ε, ο, ι> occurred much more frequently than the others. It should be noted that the frequency of the letter <η> in Samos, when compared with other regions, is unusually high. One possible reason is that the small sample size may bias the findings. Another reason is that the letter <η> in Samos might represent another vowel apart from /εː/. The real reason or reasons for the high frequency of the letter <η> remain unknown.

3. Attica

A total of eight Attic inscriptions dated the 6th century BC taken from Plate 3 (Nos. 18-21, 24-25, 28-29) in Jeffery's *The Local Scripts of Archaic Greece* were used as samples. The findings of the vowel letter frequencies are listed in Table 3 below.

Table 3 Frequencies of vowel letters in the inscriptions from Attica

α	ι	υ	ε	ο	η	ω	αι	ει	οι	ηι	ωι	υι	αυ	ευ	ου	ωυ
62	30	6	68	56	0	0	12	5	9	0	0	0	2	1	1	0
24.6	11.9	2.4	27	22.2	0	0	4.8	2	3.6	0	0	0	0.8	0.4	0.4	0

Total number of occurrences of the vowel letters: 252

It is generally believed that the Attic vocalic system of the 5th century BC comprises the following monophthongs: /a(:), i(:), u(:), e(:), o(:), ɛ:, ɔ:/. If the Attic vocalic system of the 6th century BC is the same, then the letter <ε> should represent /ɛ:/ apart from /e(:)/, and the letter <o> should represent /ɔ:/ apart from /o(:)/.

As can be seen from Table 3 above, the vowel letters <ε, α, o, ι> occurred much more frequently than the others. This could mean that among all the Greek vowels, /e(:) & ɛ:, a(:), o(:) & ɔ:, i(:)/ occurred the most frequently.

4. Crete

A total of five Cretan inscriptions dated the 7th and 6th centuries BC taken from Plate 60 (Nos. 15, 18-20, 22) in Jeffery's *The Local Scripts of Archaic Greece* were used as samples. The findings of the vowel letter frequencies are listed in Table 4 below.

Table 4 Frequencies of vowel letters in the inscriptions from Crete

α	ι	υ	ε	ο	η	ω	αι	ει	οι	ηι	ωι	υι	αυ	ευ	ου	ωυ
68	33	3	57	57	11	0	14	4	11	3	0	0	1	2	0	0
25.8	12.5	1.1	21.6	21.6	4	0	5.3	1.5	4	1.1	0	0	0.4	0.8	0	0

Total number of occurrences of the vowel letters: 264

As can be seen from Table 4 above, the vowel letters <α, ε, o, ι> occurred much more frequently than the others. This could mean that among all the Greek vowels, /a(:), e(:), o(:), i(:)/ occurred the most frequently.

The presence of the letter <η> In Crete should be noted. It is possible that <η> represented /ɛ:/ while <ε> represented /e(:)/. Should this be the case, /e(:)/ occurred much more frequently than /ɛ:/.

5. Corinth

A total of sixteen Corinthian inscriptions dated the 7th and 6th centuries BC taken from Plates 18 (Nos. 1, 3-7) and 20 (Nos. 16-20, 24-26, 28) in Jeffery's *The Local Scripts of Archaic Greece* were used as samples. The findings of the vowel letter frequencies are listed in Table 5 below.

Table 5 Frequencies of vowel letters in the inscriptions from Corinth

α	ι	υ	ε	o	η	ω	αι	ει	οι	ηι	ωι	υι	αυ	ευ	ου	ωυ
47	14	5	37	34	0	0	11	7	7	0	0	1	1	2	0	0
28.3	8.4	3	22.3	20.5	0	0	6.6	4.2	4.2	0	0	0.6	0.6	1.2	0	0

Total number of occurrences of the vowel letters: 166

No-one knows for certain how many monophthongs the Corinthian vocalic system has. If it comprises ten monophthongs, then the vowel letters <α, ι, υ, ε, o> probably represent /a(:), i(:), u(:), e(:), o(:)/ respectively.

As can be seen from Table 5 above, the vowel letters <α, ε, o> occurred much more frequently than the others. This could mean that among all the Greek vowels, /a(:), e(:), o(:)/ occurred the most frequently.

6. Sikyon

A total of eight Sikyon inscriptions dated the 7th and 6th centuries BC taken from Plate 23 (Nos. 2-4, 8, 11-13, 21) in Jeffery's *The Local Scripts of Archaic Greece* were used as samples. The findings of the vowel letter frequencies are listed in Table 6 below.

Table 6 Frequencies of vowel letters in the inscriptions from Sikyon

α	ι	υ	ε	ο	η	ω	αι	ει	οι	ηι	ωι	υι	αυ	ευ	ου	ωυ
19	14	7	25	21	0	0	7	1	7	0	0	1	0	1	4	0
17.8	13.1	6.5	23.4	19.6	0	0	6.5	0.9	6.5	0	0	0.9	0	0.9	3.7	0

Total number of occurrences of the vowel letters: 107

As can be seen from Table 6 above, the vowel letters <ε, ο, α, ι> occurred much more frequently than the others. This could mean that among all the Greek vowels, /e(:), o(:), a(:), i(:)/ occurred the most frequently.

Appendix 10
The transcription of Cantonese syllables in Arabic letters

This is how our Egyptian teacher transcribed in Arabic letters some Cantonese syllables beginning with /s/, /h/, and /tʰ/. It may be noted that for ease of perception only syllables with a high-level tone were chosen for the experiment, except the syllable 太 /tʰai/, which has a low-rising tone.

1. Syllables beginning with /s/

思 /si/ 'think'   書 /sy/ 'book'   山 /san/ 'mountain'

سِي   شِبِو   صان

2. Syllables beginning with /h/

希 /hei/ 'hope'   哈 /ha/ 'ha-ha'   圈 /hyn/ 'circle'   開 /hoi/ 'open'

حِيِ   حَا   هُون   هُوَى

3. Syllables beginning with /tʰ/

湯 /tʰɔŋ/ 'soup'   天 /tʰin/ 'sky'   太 /tʰai/ 'overly'   灘 /tʰan/ 'beach'

نُونِج   نِّن   تَلِي   طان

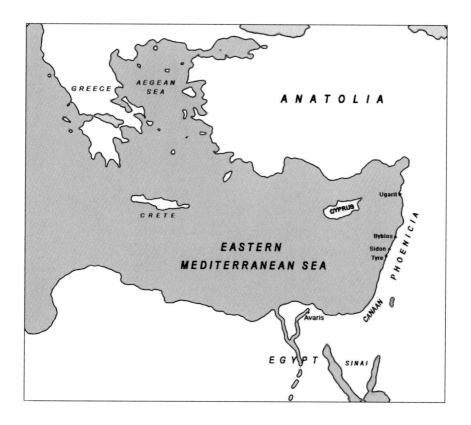

Map 1 Ancient Egypt and Phoenicia

Map 2 Ancient Greece